Unnatural Selection

Science in Society Series

Series Editor: Steve Rayner
James Martin Institute, University of Oxford

Editorial Board: Gary Kass, Anne Kerr, Melissa Leach,
Angela Liberatore, Jez Littlewood, Stan Metcalfe, Paul Nightingale,
Timothy O'Riordan, Nick Pidgeon, Ortwin Renn, Dan Sarewitz,
Andrew Webster, James Wilsdon, Steve Yearley

Business Planning for Turbulent Times
New Methods for Applying Scenarios
Edited by Rafael Ramírez, John W. Selsky and Kees van der Heijden

Democratizing Technology
Risk, Responsibility and the Regulation of Chemicals
Anne Chapman

Genomics and Society
Legal, Ethical and Social Dimensions
Edited by George Gaskell and Martin W. Bauer

Nanotechnology
Risk, Ethics and Law
Edited by Geoffrey Hunt and Michael Mehta

Unnatural Selection
The Challenges of Engineering Tomorrow's People
Edited by Peter Healey and Steve Rayner

Vaccine Anxieties
Global Science, Child Health and Society
Melissa Leach and James Fairhead

A Web of Prevention
Biological Weapons, Life Sciences and the Governance of Research
Edited by Brian Rappert and Caitrìona McLeish

Unnatural Selection

The Challenges of Engineering Tomorrow's People

EDITED BY

Peter Healey and Steve Rayner

earthscan
from Routledge

First published by Earthscan in the UK and USA in 2009

For a full list of publications please contact:
Earthscan
2 Park Square, Milton Park, Abingdon, Oxfordshire OX14 4RN
711 Third Avenue, New York, NY 10017

First issued in paperback 2014

Earthscan is an imprint of the Taylor & Francis Group, an informa business

ISBN 13: 978-1-84407-622-2 (hbk)
ISBN 13: 978-1-138-00208-1 (pbk)

Typeset by JS Typesetting Ltd, Porthcawl, Mid Glamorgan
Cover design by Susanne Harris

A catalogue record for this book is available from the British Library

Library of Congress Cataloging-in-Publication Data

Unnatural selection : the challenges of engineering tomorrow's people / [edited by] Peter Healey and Steve Rayner.
    p. ; cm.
    Based on a conference entitled "Tomorrow's People: the Challenges of Technologies for Life Extension and Enhancement" which was organized by the James Martin Institute for Science and Civilization at Oxford University's Saïd Business School and held in Oxford in 2006.
    Includes bibliographical references and index.
    ISBN 978-1-84407-622-2 (hardback)
  1. Medical innovations–Moral and ethical aspects–Congresses. 2. Rejuvenation–Moral and ethical aspects–Congresses. 3. Longevity–Moral and ethical aspects–Congresses. 4. Medical ethics–Congresses. I. Healey, Peter, 1942- II. Rayner, Steve, 1953- III. Earthscan.
    [DNLM: 1. Biomedical Enhancement–ethics–Congresses. 2. Bioethical Issues–Congresses. 3. Biomedical Enhancement–methods--Congresses. 4. Life Expectancy–trends–Congresses. 5. Longevity–ethics–Congresses. W 82 U58 2008]
    RA418.5.M4U48 2008
    174.2--dc22
                        2008036282

Contents

List of Figures and Tables

FIGURES

TABLES

List of Contributors

Bill Bainbridge, Co-director, Human-Centred Computing, National Science Foundation, Arlington, VA, USA. *wbainbri@nsf.gov*

Nick Baylis, Co-director, Well-being Institute, University of Cambridge, UK. *cambridge@nickbaylis.com*

Nick Bostrom, Professor and Director, Future of Humanity Institute, University of Oxford, UK. *nick.bostrom@philosophy.ox.ac.uk*

Robert A. Butler, President and CEO, International Longevity Center, New York, NY, USA.

Z. F. Cui, Donald Pollock Professor of Chemical Engineering and Director, Oxford Centre for Tissue Engineering and Bioprocessing, University of Oxford, UK. *zhanfeng.cui@eng.ox.ac.uk*

Aubrey de Grey, Chairman and Chief Science Officer, Methuselah Foundation, Lorton, VA, USA. *aubrey@sens.org*

Joel Garreau, Principal, The Garreau Group. *joel@garreau.com*

Robin Hanson, Department of Economics, George Mason University, Washington, DC, USA. *rhanson@gmu.edu*

Sarah Harper, Professor of Gerontology and Director of the Oxford Institute of Ageing, University of Oxford, UK. *sarah.harper@ageing.ox.ac.uk*

Peter Healey, Research Fellow, James Martin Institute for Science and Civilization, University of Oxford, UK. *Peter.Healey@sbs.ox.ac.uk*

Ellen Heber-Katz, Professor, Molecular and Cellular Oncogenesis Program, Wistar Institute, Philadelphia, PA, USA. *heberkatz@mail.wistar.org*

James Hughes, Executive Director, Institute for Ethics and Emerging Technologies, Trinity College, Hartford, CT, USA. *James.Hughes@trincoll.edu*

Rachel Hurst, Disability Awareness in Action, UK. *rachel.daa@btinternet.com*

Tom Kirkwood, Professor of Medicine and Director of the Institute of Ageing and Health, Newcastle University, UK. *tom.kirkwood@newcastle.ac.uk*

Wolfgang Lutz, Adjunct Professor of Demography and Social Statistics, University of Vienna, Austria. lutz@iiasa.ac.at

Ma Ying, Institute of Science, Technology and Society, Chinese Academy of Science and Technology for Development, People's Republic of China.

Richard A. Miller, Professor of Pathology, University of Michigan, USA. *millerr@umich.edu*

Alfred Nordmann, Professor of Philosophy and History of Science, Darmstadt Technical University, Germany. *nordmann@phil.tu-darmstadt.de*

David Nutt, Professor of Psychopharmacology and Head of Community-Based Medicine, University of Bristol, UK. *David.J.Nutt@bristol.ac.uk*

S. Jay Olshansky, Professor, School of Public Health, University of Illinois at Chicago, USA. *sjayo@uic.edu*

Daniel Perry, Executive Director, Alliance for Aging Research, Washington, DC, USA.

Steve Rayner, James Martin Professor of Science and Civilization, University of Oxford, UK. *Steve.Rayner@sbs.ox.ac.uk*

Arie Rip, University of Twente, The Netherlands. *a.rip@utwente.nl*

Anders Sandberg, James Martin Research Fellow, Future of Humanity Institute, University of Oxford, UK. *anders.sandberg@philosophy.ox.ac.uk*

Dan Sarewitz, Professor of Science and Society and Director of the Consortium for Science, Policy and Outcomes, Arizona State University, USA. *dsarewit@exchange.asu.edu*

Julian Savulescu, Uehiro Chair in Practical Ethics and Director of Uehiro Centre for Practical Ethics, University of Oxford, UK. *julian.savulescu@philosophy.ox.ac.uk*

Peter Schwartz, Co-founder and Chairman of Global Business Network, San Francisco, CA, USA.

Lee Silver, Professor of Molecular Biology and Public Policy, Woodrow Wilson School of Public and International Affairs, Princeton University, USA. *lsilver@princeton.edu*

Danielle C. Turner, Department of Psychiatry, University of Cambridge, UK. *dct23@cam.ac.uk*

Shiv Visvanathan, Dhirubhai Ambani Institute of Information and Communication Technology, Gujarat, India. *svcsds@gmail.com*

Kevin Warwick, Professor of Cybernetics, University of Reading, UK. *k.warwick@reading.ac.uk*

Gregor Wolbring, Faculty of Medicine, University of Calgary, Alberta, Canada. *gwolbrin@ucalgary.ca*

Zhao Yandong, Institute of Science, Technology and Society, Chinese Academy of Science and Technology for Development, People's Republic of China. *zhaoyd@casted.org.cn*

Acknowledgements

We would like to convey thanks to our close collaborators Demos, and especially to Paul Miller and James Wilsdon. They were a never-failing source of ideas and advice, and their edited volume *Better Humans? The Politics of Human Enhancement and Life Extension* (Demos Collection 21, 2006) contains more or less closely related pieces from a number of the authors represented here: Nick Bostrom, Rachel Hurst, Danielle Turner and Gregor Wolbring.

Susan Greenfield's book *Tomorrow's People: How 21st Century Technology is Changing the Way We Think and Feel* (Allen Lane, 2001) provided the subtitle to this volume.

The *New Yorker* cartoons in Chapter 2 appear by permission of Condé Nast Publications.

Alfred Nordmann's original presentation, on which Chapter 3 is based, is included in 'Ignorance at the heart of science? Incredible narratives of brain–machine interfaces', forthcoming in Johann S. Ach and Beate Lüttenberg (eds) *Ethics in Nanomedicine*, Lit-Verlag, Berlin. A more sustained and more circumspect development of the argument can be found in Nordmann's (2007) 'If and then: A critique of speculative nanoethics', *NanoEthics*, vol 1, pp31–46.

Chapter 6 (by James Hughes) first appeared in 2007 in Nigel M. de S. Cameron and M. Ellen Mitchell (eds) *Beyond Human Nature: The Debate over Nanotechnological Enhancement*, pp61–70, and is reprinted with permission of John Wiley and Sons, Inc.

Chapter 9 (by Sarah Harper) is an updated and abridged version of a paper published in 2006 in *Journal of Population Research*, vol 23, no 2.

Chapter 12 (by Olshansky et al) is reprinted with permission from *The Scientist*, where it was originally published in March 2006.

Parts of Chapter 18 (by Danielle Turner) were published in 2006 in *BioSocieties*, vol 1, no 1, pp113–123. Dr Turner would like to thank Barbara Sahakian, Luke Clark and Simon Redhead for helpful discussions.

Parts of Chapter 22 (by Julian Savulescu) appeared in two previous works by the same author: 'Genetic interventions and the ethics of enhancement of human beings', published in 2006 in B. Steinbock (ed) *The Oxford Handbook on Bioethics*, Oxford University Press, pp516–535; and 'In defence of procreative beneficence: Response to Parker', published in 2007 in *Journal of Medical Ethics*, vol 33, pp284–288.

The authors of Chapter 23 (Zhao Yandong and Ma Ying) thank Jon Pederson and Guihua Xie for their valuable comments on a draft of this paper.

We should like to thank our editors at Earthscan, especially Alison Kuznets and Hamish Ironside, for their skill and patience in making this book possible.

Peter Healey and Steve Rayner
Oxford
October 2008

Part I

Introduction

1

Introduction

Peter Healey and Steve Rayner

Conventional explorations of the interaction between cultural and genetic inheritance in the evolution of human beings are based on a clear line of causation: the twin tracks of that evolutionary process have shaped who we are and, in particular, shaped the religious, spiritual and moral awareness and values which many consider to be at the core of our identity, our human nature.

The chapters in this book look at what some characterize as the next stage of evolution: conscious efforts by human beings to reshape their inherited physical, cognitive and emotional identities by extending lifespan and enhancing human capacities. The values and identities go from consequence to cause – we are what we want to be.

In themselves human actions to extend lives and enhance our capacities are of course far from new, arguably dating back to the adoption of protective clothing, and include spectacles, hearing aids and joint replacements. However, the rapid advances in our knowledge in many areas of science and technology – for example genomics, stem cells, cognitive enhancing drugs, regenerative medicine, advanced prostheses, and direct interfaces between computers and the human brain – seem to suggest that something of a quite different order is becoming possible. The convergence of what individually might be seen as quite modest increments in our knowledge and capacities seems cumulatively to outrun science fiction by suggesting futures that could represent a radical discontinuity in human experience.

The origins of the book lie in a meeting in Oxford in 2006. 'Tomorrow's People: The Challenges of Technologies for Life Extension and Enhancement' was a world forum organized by the James Martin Institute for Science and Civilization (www.martininstitute.ox.ac.uk/jmi/). The Institute was founded at Oxford University's Saïd Business School in 2004, with a generous benefaction from the James Martin Trust. Its mission is to focus on the major issues of science and technology that are likely to shape the next decade to century.[1] The special role of the World Forum – and by extension the aim of this book – is to help those who engage with it, from a wide range of roles and countries, understand the range of possible futures that these will make possible and equip them to make more informed choices in managing the resulting social and technical change. In looking at issues of science and technology likely to shape the 21st century, it seemed wise for the Institute to start here, with changes in ourselves.

This book looks at the range of technologies offering lives that purport to be longer, stronger, smarter and happier, and examines under what conditions their introduction is likely to lead to more fulfilled individuals, stronger and stable communities, and a fairer world.

The emphasis throughout the book – a point that comes very clearly out of Joel Garreau's opening keynote – is that these questions do not hang abstractly over our heads, but will be resolved as the result of processes of social choice. What kind of human beings we are now, and in what variety we may develop in future, depends on our framing of both big scenarios and decisions on small issues. Selecting the timescale and scope of the issues addressed in the book and the perspectives of those addressing them – advocates of a variety of positions – was important because it itself contributes to framing those social choices.

On timeframe, we decided to look far enough ahead to present real alternative futures, but not so far that we cannot focus back on the choices that need to be made – or can be made – in the near future. Thus, by and large, we have avoided debates about whether the current flowering of the new life technologies represents part of an accelerating scientific future – a wider pattern of exponential knowledge growth – that will culminate in 'the singularity', when machine intelligence transcends that of humans, although not necessarily at the latter's expense (see Kurzweil, 2005) – although Bill Bainbridge's and Robin Hanson's chapters take us some way along this track. Similarly, we have chosen not to put the issue of potential human immortality centre stage, although the contributions to the *Longer?* section of the book show sharply divergent perspectives on what is possible, the best research approaches, and tactics in seeking public and funding support. How might the US National Institutes of Health, for example, be persuaded to recognize Richard Miller and Jay Olshansky's view that if they were to extend the proportion of their budget from the 0.06 per cent currently spent on research into the fundamental biology of the ageing process or the means to control it, the return on this investment in terms of lifespan extension would be 10 times greater than that of research designed to abolish heart attacks or cancer.

The fierce character of the debate in *Longer?* is a consequence of a third strategic commitment that we made in framing this book: in favour of pluralism – the need for the debate to reflect a divergent range of perspectives. This commitment was another consequence of our focus on the importance of exploring how values would relate to the decisions made, how far different value-sets were intrinsically conflicting and how far they might be resolved. In practice this meant making sure that the discussion not only included the usual range of professional suspects – leading researchers in the sciences on which life-enhancement technologies depend and people from the social sciences and the humanities who are skilled at assessing and communicating the social impacts of technological change – but that it also reflected different social and ideological positions that have begun to condense out of the debate.

One dimension of the debate has as its poles the bioconservatives (a label which largely originates from its opponents and which its wide-ranging adherents would not always recognize) and the transhumanists (a self-description of a grouping which has its own membership association). Bioconservatives share a belief in the 'natural' and a suspicion of technological modifications to it – seeing human nature as a given – while the transhumanists' 'most dangerous idea', as Francis Fukuyama (Fukuyama, 2004) puts it, is 'to embrace redesign of the human condition, including such parameters as the inevitably of ageing, limitations on human and artificial intellects, unchosen psychology, suffering, and our confinement to planet Earth' (World Transhumanist Association, 2002). Fukuyama sees this agenda as lacking humility and respect for human nature, and rails against the transhumanists' 'genetic bulldozers and psychotropic shopping malls', which he fears as its consequences. In Joel Garreau's overview in this book, the 'heaven' and 'hell' scenarios derive from the transhumanist and bioconservative positions respectively, while the 'prevail' alternative reflects pragmatism informed by humanism – careful and selective choice of what the new technologies offer.

There are two dimensions to the organization of this book. One dimension examines the claims of enhancement in terms of the benefits they are said to be about to bring people – lives that are longer, stronger, smarter and happier – and then goes on to ask the interrelated questions about whether such developments will be fairer, and how easily, and in what terms, they are governable. This last challenge – finding effective and legitimate mechanisms for informed public/ political choice – is particularly demanding, especially in areas of science and technology such as this where, as Bill Sharpe says, 'fact, fiction and ambition are intimately entwined' (Sharpe, 2006).

The second, cross-cutting dimension looks at some of the wider societal issues that the debate engenders: about what, in the light of the advent of our capacity to engineer ourselves, we can say about the nature of human nature, and how we might expect these new life technologies to be enacted and received in parts of the world differing in culture and wealth.

One element of these differences becomes evident in Sarah Harper's introduction to *Longer?*. She sets out the demographic expectations based on current trends – hoped-for stability in the developed countries by mid-century, with around 10 per cent of the population in each decade between birth and 100, and – with the exception of Africa, currently hollowed-out by AIDS – with life expectancy in developing and transitional regions approaching that of developed countries. Much of this is healthy life expectancy, with most of the additional years of life free of disabilities. Up to this desirable equilibrium point, however, countries have to confront an unbalanced population, with a high proportion of old and a great increase in the proportion of the oldest old, the over-80s. However relatively benign old age may become, in absolute terms inevitably this will increase the burden of disease, and as the demographic rebalancing proceeds, in the developing countries the pressures of caring for the old will combine with the challenges of reducing disease and mortality among the young. As Harper makes clear, the first requirements in responding to these challenges will be a rapid build-up in the next 40 years of the well-established life-extension technologies of clean water supply and good sanitation, technologies which the developed world has gradually built into its infrastructure over the last 150 years.

Harper's analysis highlights the bald 'development gap' of differing regional responses to the issues in this book. But equally important may be cultural and attitudinal differences, which can influence what is seen as enhancement, the priorities of research and development, and the limits to such activity which may be imposed when the limits to the society's view of the 'natural' are breached. We are reminded that the benefits of the new life technologies are not confined to their consumers, but of course will enhance the economies of those who gain an early lead.

Alfred Nordmann sets out his position in relation to differing US and European strategies to the development of the converging technologies which are often seen as the core knowledges on which the new life technologies draw (nanotechnology, genomics, cognitive systems, neuroscience, and information and communication technologies). When it comes to human engineering and technologies for the enhancement of life, the European approach to 'converging technologies' (European Commission, 2004) expresses certain core commitments. Instead of using technology to realize human potential, Nordmann sees us needing human creativity to realize the potential of enabling technologies. Rather than engineering *of* body and mind – which he implies is the approach represented in the equivalent US report (Roco and Bainbridge, 2002), we need to emphasize engineering *for* body and mind. Nordmann sees visionary speculations as a distraction from the immediate ethical and societal challenges of designing our future. If only to determine the limits of technical solutions to human and societal problems, we need to draw on basic science, the humanities and the social sciences.

Rather than expressing a specifically European perspective, Nordmann sees these core commitments as corresponding to universal principles of science. When researchers increasingly trade in images, draw on visions and depend on other disciplines, it becomes difficult even for science itself to stay within bounds of credibility and reason. Nordmann calls for watchdog activities to safeguard not only against fraudulent research but also against the self-ascribed scientific expertise of visionaries.

Shiv Visvanathan echoes and extends this debate. He invokes Karl Mannheim to characterize the European and American cultural ways of approaching a problem.

Mannheim observed that when one said the word 'problem' to a European, he treated it as a riddle, an exercise in metaphysics, a question that you unravel again and again, whereas when one suggested it to the American, he reached for his toolbox. For him, a problem was something that needed a technological solution. These two perspectives are echoed in the chapters that follow. In the debate on life-enhancing technologies, Visvanathan sees a tension between an America at one end 'bristling with technological promise and a UK and Europe seeded with philosophical doubt'.

Visvanathan's alternative South Asian perspective on these technologies is an attempt to locate them as imaginations, and to ask what imaginations of these technologies say about the link between the cosmology, constitutions and civics of these worlds: *cosmology* in the sense of relations between God, man and nature; *constitutions* in terms of the normative and legal assumptions surrounding the technologies (including notions of social contracts which may be implicit in the technologies and the ways that they are framed in policy/regulatory terms); and *civics*, the everyday way in which these technologies may be incorporated in formal explications of innovation chains. Visvanathan develops this particularly in relation both to deeper ideas of democracy, in which the public is seen not just as expressing interests, but as possessing values, ideas and knowledge frameworks critical to innovation policy, and to alternative views of the future which can inform technology assessment and forecasting. This wider view of democratic input turns forecasting from a narrow instrumental game into an unbounded process of possibilities, akin to celebratory play.

Lee Silver sees the distinct cultural perspectives of the East towards the life technologies as competitive scientific and economic advantage. From a historical perspective, he reminds us that just as biotechnology now provides the greatest hope for alleviating human suffering and, simultaneously, sustaining a vibrant biosphere, in a sense it has always done so. Human civilization was founded on the ability to control and manipulate genes in other organisms, including plants, animals and microbes; neither cows nor corn existed until human ancestors invented them. But since the detailed ancient history of this incremental innovation in plant and animal breeding is largely unknown, biotechnology appears to us as the most contentious of modern inventions, because it challenges overt, covert or subliminal Western beliefs in a 'Supreme Being' or a preordained 'Master Plan'. From conservative Christians who view embryo cloning as a violation of God's right to create each individual human soul, to New Age secularists who view plant engineering as an assault on the spirit of Mother Nature, for Silver the opponents of biotechnology are driven by faith in a higher or deeper spiritual authority who demands our allegiance. Silver contrasts this with the fluid spiritual traditions of Asian countries – where souls are eternal, self-evolving and not beholden to an external master – allowing a more ready acceptance of both embryo research and genetically modified plants, and presumably, although he does not mention this, providing a source of ready adherents to transhumanism. Because of these fundamental differences in

Western and Eastern spiritual heritage, he sees Korea, China, Singapore and India as now poised to take the lead in biotechnological advancement.

As well as having different implications for world regions, new technologies have different implications for different groups of people within society. One of the recurrent themes of this book is what life-enhancement technologies represent to those classified as disabled: opportunity or threat? Rachel Hurst traces the history of treatment of disabled people as deviant and argues that ideas of perfectability in contemporary society, coupled with denial of the natural range of human capacity and experience, provide a background against which discrimination against the disabled, up to and including death by action or neglect, can prosper. Her argument is not an in-principle one against the new life technologies, but in favour of their use to improve, rather than limit, the life chances of the disabled. Gregor Wolbring speculates as to whether an unenhanced underclass will be created. Enhancement may stretch typical boundaries of what it means to be human to the point where the technical description of human beings as members of the species *Homo sapiens* ceases to be accurate. Since each form and purpose of enhancement comes with its own sales pitches, social consequences, problems and implications, they need to be assessed on a case-by-case basis.

One of the main arguments in the enhancement debate is that you can and should make a distinction between therapy and enhancement. However, this argument and many others employed in the enhancement debate depend on what concept of health you follow. Wolbring concludes that the key question is which concepts of health, disease, disability, wellbeing and even medicine we use. So we return to the evolving values which shape our choices.

If we consider the values which are shaping our current choices in the development of life-enhancement technologies, one area that deserves further exploration is the implications of the use of human enhancement by the military. We know that the military, especially in the US, are sponsoring research on enhancements of potential value on the battlefield. These include exoskeletons to allow the soldier near superhero strength in moving or carrying, suits that offer built-in immediate medical diagnosis and treatment of the wounded in the field, and for amputees, limb regeneration. Will such developments, many currently being pursued through MIT's Institute for Soldier Nanotechnologies,[2] increase the relative strength of the dominant military hegemonies which are likely to command them first?

There may be implications of human life extension and enhancement not only for the conduct of military conflict, but for its likelihood. It may be that the overwhelming problems of demographic imbalance resulting from survival of the aged lead to national introversion, sapping motivations for war and pre-empting the resources for them – a largely pacific effect. Alternatively the problems of imbalance may increase the attraction of population 're-balancing' through 'acquisitions' of younger populations through migration and political 'mergers' – processes that may spark social conflicts within or between societies. The enhanced consumption of the physically and sensorially enhanced will – if nano fails to deliver on its hype of cheap and abundant goods delivered in a clean environment – intensify the competition for natural resources. If this leads in turn to new conflicts, which nations and which individuals within each nation (will we see conscription for the old?) will be involved?

As if to indicate the scope of the choices that will shape the technology that will shape ourselves, the chapters of this book reflect a wide range of views as to what will be possible, on the likely timescale of major developments, and on the scientific approaches that will yield results, their desirability and their social impacts. Thus the transhumanists believe that our inherited capacities are an arbitrary set of what is possible and it is imperative to explore and take advantage of every option to enhance our senses and to expand our cognitive and our physical selves (before the enhanced ultimately transcend physical form by becoming pure information). For others, in an echo of the debates over eugenics in the 1930s, the attempt to build such superhumans suggest a huge and growing divide in humanity in the longer term, and more immediately more negative views towards the disabled. For some, the focus is on understanding the mechanisms of ageing; for others it is cleaning up the damage that accumulates in the cell with age. For some, the 'longevity dividend' is a strategic goal of a 1000-year lifespan (albeit achieved in incremental steps which keep the individual just ahead of the grim reaper – what de Grey calls 'death escape velocity'); for others it is using our new understanding to defeat the diseases that often make longer life a misery.

In practice, once the first steps in the development of these technologies become established, consumer action may set a critical path through these competing narratives. Members of a new wedge of 'healthy survivors' may become advocates of further capacity enhancements. Along with these healthy survivors we may see the survival of another wedge of the most vulnerable, with consequent increased demands on family/institutional carers. Will those whose life is unexpectedly extended have qualitatively different experiences and values, and consequently want to make different social contributions? We could see the start of a social dynamic that might drive the development of biological speciation – the separation of the enhanced from the unenhanced – and the pressure for further enhancements to be technically – and legally – available. Consumerism and regulation may find themselves at odds, particularly in areas of traditionally strong governance, such as the uses of drugs, those presenting novel ethical issues, or novel technologies where the distribution of risks and benefits seem hard to assess. Differences in viewpoint on these issues – such as we have seen in the US with stems cells, with bioconservatives, pragmatics and transhumanists taking different lines – may congeal into the more permanent divisions of a new biopolitics, with diverging definitions of what it is to be human opening up between countries as well as within them.

The threads of discussion, and suggestions as to how they may be resolved, are brought together at the end of the book by articles on regulatory issues by Arie Rip and Dan Sarewitz, and a view from the future, 42 years hence, by Peter Schwartz, a distinguished builder of scenarios and writer about the future who is also the chairman of the Global Business Network. For Rip the key term is technology 'modulation' – modification of technologies through the heterogeneous networks of science, industry, policymakers and societal stakeholders which shape them. 'Modulation' in this sense – making such networks reflexive and anticipatory, so that they can be more intelligent in enabling and constraining future developments, is the peak of Rip's ambition for regulation, and is contrasted with what he sees as the illusion of central control.

Sarewitz echoes Rip's view of regulation as a process of governance – in which outcomes are determined through interacting networks of actors, rather than by one focal institution. For him, the barriers separating right from wrong are fuzzy, somewhat arbitrary, often strongly culturally determined and yet likely to evolve over time. Far from these qualities rendering these boundaries valueless, for Sarewitz they are essential to making sense of the world and building rule systems to keep chaos at bay. For this and other reasons, he expresses irritation with the transhumanist view that since humans have long used enhancements, all future such developments are to be welcomed.

Peter Schwartz adopts the entertaining and enlightening technique of addressing us – as an enhanced 104 year old – from the future of 2050. He portrays a world of happier, healthier people living in a sustainable ecosystem in a highly productive economy, but one in which there had been disasters en route, where different routes to the future were taken and where different capacities resulted.

We see the book's legacy as two-fold. First, we hope to be influential in setting an agenda for future public debate about the issues that surround the development and introduction of these new technologies. Second, and more ambitiously, we see the opportunity for new institutional forms – new coalitions between academia, business and civil society – to explore the ground for innovation that is both technically and socially robust, in that it is seen to respond to the social and ethical questions posed both within advanced societies and in relation to disadvantaged parts of the world. Our Institute sees the development and assessment of these new approaches to scientific governance as one of its continuing roles and looks to find partners to play a major role in scoping and developing them.

NOTES

1   A further benefaction from the Trust in 2005 established Oxford's School of the Twenty-First Century (www.21school.ox.ac.uk).
2   http://web.mit.edu/isn.

2

Radical Evolution: An Overview of the Near Future

Joel Garreau

I'm not all that interested in technology. I'm interested in culture and values – who we are, how we got that way, where we're headed and what makes us tick. I'm interested in the future of human nature.

We are at a turning point in history. For the first time in hundreds of thousands of years, our technologies are not so much aimed outward at modifying our environment in the fashion of fire and clothes and agriculture and cities. Instead, our technologies are increasingly aimed inward – at modifying our minds, memories, metabolisms, personalities and progeny. If we can do all that, the question is whether we can change human nature and whether this is going to be happening on our watch. Not in some distant science-fiction future but right now – in our generation. We're the first species to take control of our own evolution. I call this 'radical evolution', and what I want to present is a road map for the next 10 or 20 years – the near future.

In the very near future your child is going to come home in tears, again. The reason will be that he once again finds himself unable to compete with children who are more athletic than he is, smarter, better behaved, more beautiful, more capable, and certainly better at getting into the best colleges, because their parents invested in enhancements for their child, and you did not. The question is: what are you going to do when this day occurs?

There are several possibilities. One is that you say to him, 'I don't care what other parents do to their kids' minds and bodies. We love you just the way you are. Wipe your nose.' Another possibility is that you go out and remortgage the house again in order to compete in this arms race with other parents who are enhancing their children. And the third possibility is that you could try to get these enhanced

"Obviously, they're evolving much faster than we are."

Figure 2.1 *Palaeolithic ping-pong players*

Source: The New Yorker (reproduced by permission of Condé Nast Publications)

children thrown out of your schools. The trouble with that is it just widens the gap between people who are enhanced and people who are 'naturals', if you will. What does that do to our culture and values? Does that create even larger problems than we've had before?

But what you can't do is ignore this because these changes in what it means to be human are coming at us right now. Let me tell you how all this came about.

When lecturing I quite often ask who knows what I mean by 'Moore's Law'.[1] One of the things to keep in mind as we head into this future is that when I ask this question, even among the most intelligent, sophisticated business people, I'm lucky to get a 5 per cent or 10 per cent show of hands. So the first thing that those of us discussing radical evolution need to do is start working on our communication skills because an awful lot of what we take for granted is not in the public domain.

We are very conscious of technical change. Many are old enough to remember when computer screens were only black and white, or rotary dial phones, or the smell of mimeograph machine fluid, or even polio. The reason so many things that we can recall are no longer here is that we are going through a curve of exponential change. You can think of it as a third wave of evolution.

The first wave is biological evolution – what Darwin looked at. In this wave it took hundreds of millions of years to get from bacteria to vertebrates, and then 150 million years to get from the first mammals to the first monkeys, then 30 million years to get from monkeys to chimpanzees, then 16 million years to get from chimps to walking erect, and then 4 million years to get from walking erect to painting on cave walls, and then 10,000 years to get from painting on cave walls to fixed settlements. So you can see that the development of cognitive function has followed an accelerating curve, with intelligent life emerging only very recently.

When we got to fixed settlements and invented writing so we could store and share our thoughts, that's when we hit a second kind of evolution: cultural evolution. That second wave of evolution took 4000 years to get to the Roman Empire, then 1800 years to get from the Roman Empire to the industrial revolution, and then 100 years to get to the first flight. After that, from the first powered flight to walking on the moon took only 66 years. So you can see that cultural evolution also shows a curve of exponential change.

Now we have entered the era of Moore's Law. The amount of computer firepower you can buy for a dollar has doubled 31 times since the chip was first introduced in 1959. That's an increase of over a billion times in the lifetimes of many people. We've never seen anything like this curve of change in human history. There are now cellphones with more computer firepower than was possessed by the entire North American Air Defense Command at the time Gordon Moore first prophesied his law in 1965. This exponential growth in computer power has brought us to the third kind of evolution – radical evolution – where we become the first species to take direct control of our own evolution.

One of the things to keep in mind about the arithmetic of Moore's Law is that it means that the last 20 years is not a guide to the next 20 years. If you're doubling every 18 months, it's at best a guide to the next 8 years. Similarly, the last 50 years is not a guide to the next 50 years. By this arithmetic it's at most a guide to the next 14. This is one of the reasons why we have to start focusing on how we are going to co-evolve with these challenges. I think we've got a long way to go and a short time to get there.

I'm not going to make any predictions about the future, but what I can examine here is what's in the labs and what's in existence and talk about the implications for our culture and values.

For example: Belle is a telekinetic monkey. She's at Duke University. She's an owl monkey. And she can move objects long-distance with her thoughts. How do you make a monkey telekinetic? For the better part of a year I hung out at DARPA, which is the Defense Advanced Research Projects Agency in Arlington, Virginia. They're funding technology with which they got Belle – who is a very clever little monkey – hooked on computer games. She had a joystick; if she took her cursor and hit a flashing light, she got a drop of juice. But within a matter of days it became clear she did not care about the juice. She was hooked on the game.

Then the researchers at Duke drilled a hole in her head. They took a device about the size of a baby aspirin that had hundreds of super-fine wires coming out of it, and they put this in her motor cortex, which controls her muscle movement. They lined up these wires with individual neurons of her brain so they could tell what individual neurons were doing.

They sealed her up, fired up the computer game, and as she played the game they watched her neurons fire. They figured out the pattern of what was going on in her brain. Then they disconnected her joystick, so there was no connection between her arm and the computer. They fired up the computer game and she was playing entirely with her mind, moving the cursor. Then they took the signal and piped it 600 miles north to MIT where there was a robotic arm that danced like a ballerina just the way her arm would have moved had she been moving her arm to play the game. If you've got a monkey that can control a robotic arm long-distance, with her thoughts, I submit to you that you've got a telekinetic monkey.

Now, the obvious question is why would you want to do this to a monkey? Good question. And, as usual, there are two answers. There is the official answer, and then there is the real answer.

The official answer is that an F22 jet fighter is a very difficult machine to operate with a joystick. How much better, DARPA thinks, if you could create a direct connection between your mind and this machine? So that not only could you control the machine with your thoughts – you think about how you want it to fly and it flies – but also you'd have the feedback loop such that everything that this machine can sense – radar, radio, heat, you name it – could be piped directly into your brain in a fashion that bypasses screens and keyboards. That would be a connection between us and our inventions that would blur the distinction between the made and the born. That's the official reason.

The real reason is that the person who was running this operation, Michael Goldblatt, has a daughter named Gina Marie. Gina Marie is a very talented young lady. She's at the University of Arizona and speaks multiple languages. She also has cerebral palsy, so she is supposed to spend the rest of her life in a wheelchair. Michael Goldblatt is completely upfront about how he has been spending untold millions of taxpayers' dollars to get his daughter out of that wheelchair. In principle, if you can control a machine with your mind, there is no practical reason why those machines couldn't be in Gina Marie's legs and let her get up and walk.

This is an important aspect because, again, we're talking about human nature, and whether we can change what it means. These are questions that are raging around the world right now in a very primitive way. Take, for example, Barry Bonds: he is a famous baseball slugger who's broken all sorts of records. It turns out he's implicated with the steroid trade and human growth hormone. There is a raging question in the sports pages right now about whether he's the same kind of human as the people whose records he broke. Should this guy go around with an asterisk on his forehead for the rest of his life marking him as a different kind of human being?

Take another example: Matthew Nagle, who was stabbed in 2001, had his spinal cord severed at the neck, so he was paralysed from the neck down (in 2007 he died of his injuries). In the summer of 2004, he had the same kind of technology implanted as Belle, the telekinetic monkey. Using this technology, he became the first human to send an email with his thoughts. What did that make him? Was he just an academic curiosity, or are we crossing any lines?

A quick review of the technologies that are driving this change will show you the power of what we're talking about. There are four technologies that are driving this radical evolution. I call them the GRIN technologies: genetics, robotics, information and nanotech.

The genetic or biological technology is all about manipulating and understanding cells at the most basic level. This has major near-term outcomes. Right now there are four US companies in stage two clinical trials on memory pills. That means you could have a commercial product in three years. All of these advances follow the same pattern. They're usually aimed first at the sick – in the case of memory pills, people with Alzheimer's. Then they move out to the needy well – Baby Boomers trying to banish their senior moments. Then they move out to the merely ambitious – like schoolchildren. Educational Testing Service (ETS) – the people who run the tests in the States that determine your ability to get into college – are beginning to realize that with these memory pills you could possibly buy an extra couple of hundred points on your childrens' SATs. What happens when that starts?

There are so many more changes in the pipeline concerning what it means to be human. Sleep is another basic human activity that's being investigated. Whales, for example, don't sleep – at least, not the way we do. Because they're mammals, if they slept like us they'd drown. DARPA is very interested in this, and it is funding an awful lot of technology that is meant to direct brain activity so that you can cope without something as basic as sleep.

Hunger is another activity that DARPA is funding research into. They're thinking that instead of sending in soldiers lean and mean, perhaps they should send them in mean and plump. Right now you just can't pump enough calories into a Special Forces soldier to keep him going. How much better it would be, they think, if we could access at will the energy that's stored in our fat.

Figure 2.2 is an example of robotic technology – the R in the GRIN technologies. Robotics refers to the melding of the made and the born: the smart machines that are doing our bidding. In 2002, one of these Predators in Yemen had a Hellfire missile under its wing, and it was tracking an SUV full of Al Qaeda leaders; the missile incinerated the SUV, so the Predator thus became arguably the first homicidal robot. We can argue whether or not this thing was really a robot – because there was after all a human in the loop, deciding to pull the trigger. But such distinctions might be viewed as overly fine by the people who were in the SUV.

Figure 2.2 *MQ-1L Predator UAV armed with AGM-114 Hellfire missiles*

Source: USAF photo via public domain website *www.vectorsite.net/twuav_13.html#m2*

The point is that if you can follow a car chase in Yemen from a desk in Miami you have basically got one of the powers of our superheroes from the 1930s and 1940s – telescopic vision. Every single one of these superpowers from our comic books is now either in existence or in development. There is an exoskeleton suit at Berkeley – which I've jitterbugged in – that allows you to lift 180 pounds as if it was 4 pounds. Its inventors are very interested in trying to figure out whether this wearable robot suit would allow you to leap tall buildings with a single bound. Actually, leaping the tall buildings is not the hard part; they know how to do that. It's the landing that they're still working on. There's a lot of this that's still in development, but already there are functional MRIs that see brains working in real time, holding out the promise to tell truth from lies, invoking the powers of The Shadow – who knew what evil lurks in the hearts of men. That power might be available to us in the very near future.

Figure 2.3 shows the I in the GRIN technologies: information technology. Here is the real personal computer of our age – the mobile phone. Because of Moore's Law, we're increasingly seeing the reduction in scale and price of everything. Not long ago, a music player used to be something that you put on your shelf. Then it became something that you carried, like the Walkman. Now it's jewellery, like an iPod Nano. Pretty soon it's going to be an earring, and this earring will very soon be a universal translator. In Afghanistan they already have a universal translator

Figure 2.3 *Mobile telephone*

about the size of a book that troops wear on their chest. There is no reason that this couldn't be something that goes right into your brain. Already there are cochlear implants and retinal implants. These are computers that stick into your ear or into your eye that are direct connections to your brain.

The last of these GRIN technologies is nanotechnology. This is the notion that you can build anything you want one atom at a time, or one molecule at a time. There has been an estimate that as many as half of the Fortune 500 companies might not be in existence in ten years if they miss the curve on nanotechnology. This technology is expected to be an industry worth US$1 trillion a year – the size of the GNP of Canada – by the year 2015, according to a US Congress study. The implications are amazing. The bigger your problem, the more attractive nano is. Take the energy crisis, for example: there is a firm that we at Global Business Network have been watching very closely called Nanosolar, which is on the verge of starting its manufacturing process to create nano solar cells. That's a small example, but that alone could have enormous implications just in the next ten years.

What are the GRIN technologies going to mean to who we are as human beings? What are they going to mean to our civilization in our lifetime, in this next generation? What I found as I examined this was that there were three scenarios

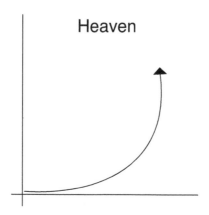

Figure 2.4 *The Heaven Scenario*

for the next 20 years. A scenario is not a prediction; it is a credible story that fits all the facts that can describe how our futures may go. The trick is to look at all these scenarios seriously. Any one of these might turn out to be the case, or it could be a combination of them. The first one is The Heaven Scenario.

Ray Kurzweil is the poster child for The Heaven Scenario. This is a gentleman who is a famous inventor and entrepreneur in the States. He's written a book called *The Singularity is Near* (Kurzweil, 2005). Ray believes that if you can live for another 20 years, the curve of this technology will be catching up in a fashion such that if you've got a good health plan and the technology is moving faster than you're ageing, you're effectively immortal.

There are many people who are actively engaging in attempting to create The Heaven Scenario today. What it essentially encompasses is the ability to conquer stupidity, ignorance, ugliness, disease, pain, suffering, obesity, forgetfulness – and possibly even death. Already we see quotes in the media about 40 being the new 30, and who hasn't talked to a grandmother who is going off to walk the Great Wall of China? There are scientists at the National Institutes of Health in Bethesda who have got a bet that the first person to live robustly to the age of 150 is already alive today. These could be early examples of The Heaven Scenario, in which the curve goes straight up, and everything is great.

That, of course, is not the only possibility. There is a mirror image possibility, which is The Hell Scenario, and again there are all sorts of distinguished explicators of how this could go. To oversimplify slightly: the people who look at The Hell Scenario see the same curve of technological change as The Heaven Scenario, but they say, 'wait a minute, this could go just the other way. The power of these technologies are amazing. What happens if they get into the hands of bumblers or fools or psychopaths?'

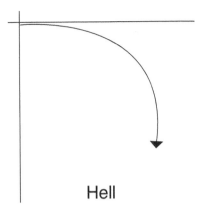

Hell

Figure 2.5 *The Hell Scenario*

The famous example is the Australian mouse pox incident. Australia is an isolated ecosystem where there have occasionally been introduced species that simply run amok. Mice are one of them: from time to time it's just mice everywhere, so they're very keen on controlling them. They were looking for a mouse birth-control device. Instead they created a monster – they took a mouse pox virus and they made one genetic change. All of a sudden, this virus was 100 per cent fatal. Every mouse died. After creating this fatal virus, they posted their methods on the internet – easily accessible to anyone who wants to replicate it.

This just drives people like Bill Joy – who's one of the more renowned Hell Scenario advocates – crazy. His view is that although a mouse pox virus doesn't hurt humans, it's a very close relative of smallpox, which obviously does. This technology is not all that complicated – any reasonably talented graduate student in a well-equipped lab could do this sort of manipulation.

One of the big issues that concerns people who talk about The Hell Scenario is this: any time you've got a self-replicating anything – whether it's a biological virus or a computer virus or a nano robot or whatever – you'd better know where the off switch is. Because if it keeps on replicating, the optimistic version of The Hell Scenario is that we simply wipe out the human race within 25 years. The pessimistic version is the one in which we wipe out all of life on Earth.

But there are a couple of problems with both The Heaven and Hell Scenarios. One of them is that the people who advocate them usually don't see them as scenarios that are mere possibilities. They see them as predictions. That's a problem because it means that the Heaven and Hell advocates talk past each other; they don't listen to each other. They tend to think that the other person is an idiot for not agreeing. If you only focus on one of these scenarios to the exclusion of all the others, you might see all of the problems but you might not see the opportunities. Or vice versa.

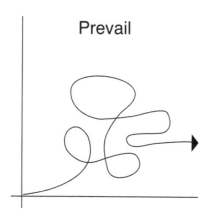

Figure 2.6 *The Prevail Scenario*

This talking past each other is why I got very interested in the third scenario, which is Prevail. I think it is a much more human version of history. Human history has very rarely followed nice smooth curves. The problem with Heaven and Hell is that they are techno-deterministic. They're visions of the future in which the technology shapes our history in an almost inevitable fashion. But human history doesn't work that way so The Heaven and Hell Scenarios don't make convincing human stories. It's just not that great a summer movie: the world is changing beneath our feet, there's not much we can do about it, hold on tight, the end. Gangbuster special effects – fabulous movie in that regard, but not much in the way of plot. Whereas the stories we tell about the future of human things usually have many more reverses and belches and loops as we kind of muddle through.

And that's the heart of The Prevail Scenario. It's not a scenario that's some middle ground between Heaven and Hell. The Heaven and Hell Scenarios are based on the notion that what matters is how many transistors you can get to talk to each other, because that's what drives this curve of exponential change. Prevail fundamentally believes something else – which is that what's critical in the development of human history in the past has not been how much technology you can get to talk to other technology but how many perverse, creative, intelligent, unpredictable human beings you can get to talk to each other.

Prevail is about the humans, not about the technology. It's about how the humans are hooked up. The significance of Prevail is that the question we are asking is 'are we going to co-evolve with our challenges?'. Because if you accept this curve of exponential change and you see the challenges coming at us on the same upward curve, and if our response is more or less flat – we're doomed.

The critical question in Prevail is: are we going to invent new social forms and new social agreements in a fashion that more or less matches our challenges? Are we going to be seeing *two* curves going up? Are we going to see co-evolution?

An example in history might have been the European Middle Ages, which were characterized by pestilence, death, violence and marauding gangs. You can well imagine why people were not particularly optimistic about the future of the human enterprise in the 1200s. Then, in 1450, there was the printing press. All of a sudden people could share and store and transport human thoughts in ways that had never been possible before. The end results were reactions that were beyond the imagination of any one human being, or even any one country. Next came global trade, the Renaissance and the Enlightenment, which led to democracy and science itself.

Can we imagine a circumstance in which we humans – newly connected through our technologies – will invent new forms of co-evolution equivalent to the ones that we've seen in the past? There are some reasons for guarded optimism

"I was wondering when you'd notice there's lots more steps."

Figure 2.7 *Lots more steps*

Source: The New Yorker (reproduced by permission of Condé Nast Publications)

in this regard. On 11 September 2001, the hijacked aeroplane United Airlines Flight 93 never made it to its target. What happened was that several dozen people on board that airplane, empowered by their airphone technology, figured out, diagnosed and cured their society's ills in under an hour flat. Was it an ideal solution? No, they paid the ultimate price. But it was good enough, and I have a hunch that that's what this kind of co-evolution looks like.

This kind of co-evolution involves bottom-up solutions. It's not waiting for anybody at Harvard or Oxford to tell them what to do. The Flight 93 passengers figured it out in a human way from the bottom up. That's the essence of Prevail: it's basically a bet on humans being surprising. It's a faith in humans coming together in unprecedented ways to be unpredictably clever. There was a conservative commentator in the 1970s named William F. Buckley, with whom I agreed on almost nothing, the exception being that he once said he would rather be ruled by the first one thousand names in the Boston phone book than by the entire faculty of Harvard.

Prevail is a bet on human nature. It's a bet on the notion that even as our circumstances change, our ability to surprise and come up with new solutions will in fact not stop. That our futures are not constrained by some inexorable curve over which we have no control. It's a bet that we are looking at the ultimate final exam here; that the technologies that we're looking at can create the Hell of a total wipeout, but it can also create transcendence. It can also dramatically change what we are capable of.

The significance of all this is not just that we're talking about all of our jobs and what it will take to compete, although that is part of it. It's not just that we're talking about all of our investments and how we will survive, although that is part of it too. And it's not just that we're talking about how gracefully all of us will age and whether we'll have many more years in front of us than we might have expected, although that too is the case. The ultimate reason we're asking ourselves these questions is that, for us, and for our children, and for our children's children for generations to come, what we are creating are new definitions of what it's going to mean to be human.

NOTES

1   Moore's Law refers to a prediction, originally made in 1965 by Gordon Moore, about the exponential increase in computing power relative to cost. Originally formulated as a doubling every year, Moore later (1975) adjusted the projection to a doubling every two years.

Part II

One World or Several?

3

Tomorrow's People, Today's Challenges

Alfred Nordmann

Today's people are different from yesterday's; tomorrow's people will be different from today's. These changes come with what we read, think and write (culture), how we organize our relationships among people (politics), how things get integrated into our lives (technology), and the multiple interactions between all this and more.

There are fascinating stories to be told about how today's people became different from yesterday's. These stories do not suggest that we were ever able to plan or foresee the effects of historical and technical change upon ourselves and who we are. Technology always helped us solve problems while creating problems of its own; we expanded our powers here and had to conform there. As we gradually came to terms with new devices or routines and assigned to them a definite place in our world, we more or less subtly transformed ourselves along with our technology, culture and social organizations.

As for tomorrow's people, there are no stories to be told as of yet, only interesting questions to be asked. These are among the more urgent:

- What can we know about tomorrow's people (tomorrow's culture, politics, technology)?
- How, if at all, might we (and should we) influence what tomorrow's people will be like?
- If all we have are more or less improbable scenarios of technological development, at what point do we need to take those seriously?

These questions don't often get asked explicitly, but by answering them implicitly we can gain or lose credibility; we can engage societal issues responsibly or irresponsibly (compare Grunwald, 2006). We might therefore explore how credibility and responsibility are at stake as we adopt one of the following three approaches to the subject of 'tomorrow's people' – where the second and third appear to correspond to the 'world regional perspectives' of the US and Europe respectively:

1   we can use visions of tomorrow's people to ask timeless philosophical questions;
2   we can foreground tomorrow's people as we address what science and technology have in store for us – for example the US initiative NBIC Converging Technologies for Improving Human Performance (where NBIC stands for nano-bio-info-cogno technologies); or
3   we can seek to enable tomorrow's solutions for today's problems, aware that these solutions create new problems for tomorrow's people – for example the European conception CTEKS (Converging Technologies for the European Knowledge Society).

In the end, however, the difference between the second and third approaches is more fundamental than that between world regional perspectives. Quite simply, the second approach must be rejected, and only the third is compatible with the traditions and values of both the US and Europe.

PHILOSOPHY

Our philosophical interest in the question of human nature provides a first context for our interest in tomorrow's people. Indeed, if we seek to understand ourselves, there is hardly a more urgent question to ask than 'Suppose you were free to choose your body and mind, would you choose yourself more or less as you are?'. We can put this question more urgently by suggesting that there might actually be specific means of remaking ourselves – ways to surgically beautify or chemically dope ourselves, and schemes of becoming immortal, controlling machinery by thought alone, or enhancing our abilities to acquire and process information.

These prospects make for fascinating, indeed endless discussions, and they do so quite irrespective of any particular beliefs or commitments about the future. Indeed, if science fiction scenarios allow us to ask interesting philosophical questions, it is precisely because we suspend disbelief in the presence of fiction. Relieved of the pressure to determine what is true or false, what is likely to happen and what not, we can forge ahead and explore who we are, who we might wish to be and how these wishes reflect on our views of human nature.

Philosophers are notorious for using improbable scenarios in order to press the issues. They take them seriously enough to generate insights from them and discover values that might guide our decisions regarding the future. But they do not take them seriously enough to believe them. As the chemist George Whitesides pointed out in regard to current discussions, when we refer to emerging science and technology in order to question some of our most basic assumptions, we must refrain from making predictions that we cannot support (Whitesides, 2004).

CONFRONTING THE FUTURE

Nowadays, the question of tomorrow's people is typically raised in quite a different manner. We are told that we must deal with this question because of current developments in science and technology. If we want to be prepared for ethical challenges that lie ahead, we have to countenance emerging capabilities of human enhancement, life extension, brain–machine interfaces and the like.

This way of questioning clearly makes assumptions about the future and how we will get there. It advances claims that demand our attention. And while it engages us in the pros and cons of certain issues, it does not allow us to question their supposedly factual foundation. Indeed, cynics might suspect that institutions like the Center for Responsible Nanotechnology or the Institute for Ethics and Emerging Technologies use ethical concern for precisely this rhetorical effect, namely to draw us into underlying visions of molecular manufacturing or a transhumanist future ('if it has societal impacts, it must be real'). To this list of institutions can be added the Foresight Institute, some publications by promoters of the US Nanotechnology Initiative and, especially, the NBIC initiative for improving human performance (Roco and Bainbridge, 2002).

As we enter into this line of questioning, we need to buy into claims about the future. It would seem all the more important, therefore, that its underlying assumptions are well founded, and it is all the more disconcerting that they are speculative at best, brazenly unconcerned with reality at worst. Three illustrations must suffice here.

First, there is Nick Bostrom's ingenious proposal to shift the burden of proof, suggesting that we are taking a big risk when we don't take seriously what may not happen for thousands and thousands of years, if ever:

> *To assume that artificial intelligence is impossible or will take thousands of years to develop seems at least as unwarranted as to make the opposite assumption. At a minimum, we must acknowledge that any scenario about what the world will be like in 2050 that postulates the absence of human-level artificial intelligence is making a big assumption that could well turn out to be false. It is therefore important to consider the alternative possibility: that intelligent machines will be built within 50 years.* (Bostrom, 2006, p41)

Our state of ignorance about the future is here taken to imply equiprobablity: if we can't be sure that something is impossible, it may just as well be likely. Instead of seeking better information and instead of focusing on the consequences of currently funded research visions, we thus find ourselves contemplating the ethical and societal consequences of a remote and hypothetical future.

Another, far cruder, way of drawing us into hypothetical visions of the future imposes a simplistic template upon the past and extrapolates from it. The most popular such template – and the least credible among historians – is a curve of exponential growth that shows how the development of technology has been accelerating throughout history and 'therefore' will continue to do so. Rather than ask whether such a 'historical law' is even remotely plausible, we are challenged to imagine what technologies will deliver such growth rates into the future.

Just two observations must here take the place of a more detailed critique of the flawed methodology behind this template. The curve of exponential growth transfers to the whole of technological development a 'law' (Moore's Law) that for the last few decades has functioned as a road map in the semi-conductor industry and which has everyone guessing how much longer it can be upheld. Also, we might just ask ourselves whether a person who was born in 1920 and died in 2000 really witnessed a faster rate of technical development (and not just a faster turnover of products) than a person who also lived for 80 years, but from 1880 (before the telephone, automobile and electric lighting) to 1960 (in the age of space travel, mass communication and transportation, and nuclear power).

Finally, if the notion of exponential growth makes a very strong claim about the history of technology, others try to get away with extremely weak claims. They extrapolate from the past to the future by making all technology look alike. John Harris and Arthur Caplan, for example, maintain that all technology serves the creation of tomorrow's people who enhance their capacities:

> *That's what agriculture is. That's what plumbing is. That's what clothes are. That's what transportation systems are. They are all attempts by us to transcend our nature. Do they make us less human?* (Caplan, 2006, p39)

According to these arguments, we are either for the technologically enhanced individuals of tomorrow or we are against all technology and the very idea of human betterment. But Caplan's offhand remark is plainly wrong on three counts. First, agriculture, plumbing and transportation systems did not seek to overcome or transcend our nature. Instead of expanding our physical and cognitive limits, these technologies attempt to render the world more manageable for creatures with our limited physical and cognitive means. Second, to the extent that agriculture, plumbing and transportation systems shape the world we live in, our values and identities are changing alongside these technologies. But surely this is not change in the direction of transcendence. As we render the world more habitable for creatures like ourselves, we do not just liberate ourselves and extend power, but also create new dependencies, new kinds of ignorance, new problems even of human or ecological survival. And finally, Caplan's remark conflates a transhumanist interest in technology for individualized human enhancement with the tradition of enhancing ourselves through education and ingenuity. He neglects to mention that currently popular visions of human enhancement, molecular manufacturing and global abundance do not aim to continue the tradition of technologically cultivated life-forms or public infrastructures (agriculture, plumbing, transportation systems), but seek to individually liberate us from the need to use our native intelligence in order to get the most out of limited resources.

Consider, in contrast to all of this, the case of global warming. Here, we have been careful to convince ourselves that the trend is real or at least highly plausible – there are no assumptions about entirely new capabilities emerging or about sudden leaps in basic scientific knowledge, there is no denial of technical or scientific limits, there is no blanket appeal to supposedly unfathomable deliverances of nano- and biotechnologies and their convergence. Also, after we persuaded ourselves of the real danger of global warming, we started looking for cultural and political interventions along with highly specific technical programmes that might slow down or perhaps even stop the trend. We do not sit down to develop an ethic for a futuristic world-under-water.

Fixated upon unlikely future scenarios of technologically enhanced individuals, we may actually blind ourselves to the transformative potential of current technical developments. Global warming is one of these, the creation of smart environments through ubiquitous computing technologies is another. Against the prospect of life in a greenhouse, with memory and intelligence incorporated into the environment, the notion of an individual human being with a memory-enhancing brain implant appears not only less likely but also pathetically irrelevant and touchingly old-fashioned.[1]

CONFRONTING THE PRESENT

If we want to reflect upon human nature, let us stimulate debate by imagining cyborgs, human consciousness uploaded into machines, or the achievement of immortality. However, if we want to consider scientific and technological developments, let's steer away from unfettered imaginings. Instead, we should look at the problems that are facing us today, problems like global warming, water supply, obesity, illiteracy, energy generation and storage, and ageing societies. And for these we need to mobilize human ingenuity. Rather than assume that technology follows a single trajectory (that of expanding human powers or realizing human potential), we should assume that there are many trajectories as we choose to draw upon and realize the potential of technology. And just to play it safe, we will not be looking for technological solutions alone.

This contrast applies to the US report on NBIC convergence for improving human performance and its European counterpart CTEKS (HLEG, 2004).[2] CTEKS involve a process of agenda-setting that matches technical capability with social imagination. Of course, visions of human enhancement might enter into this agenda-setting process. But how likely is it that immortality, thought-controlled machines or individualistic enhancements of memory will rank highly among the needs of ageing societies in a world of scarce resources?[3] Therefore, if the CTEKS report rejects engineering *of* the body or *of* the mind in favour of engineering *for* body and mind, this owes not primarily to ethical conservativism or considerations of technical feasibility. Rather it reflects the fact that engineering *of* body and mind represents an inefficient and unoriginal use of technology.

To be sure, as we mobilize scientific and technological resources to address urgent problems of today, this will have consequences for tomorrow's people. Even if we cannot foresee these consequences, they need to be a matter of concern and may even serve to motivate a responsive engineering practice. If, as James Martin suggested, we will better manage a planet with scarce resources by spreading literacy worldwide (and using cheap computers to do so), there is hardly a more positive vision for the creation of 'tomorrow's people'.

BROAD COMMITMENTS

By sharply rejecting the second approach (confronting a merely imagined future) and endorsing instead a process of agenda-setting for technological development, I appear to be endorsing a European as opposed to an American approach: instead of individualistic consumerism, I recommend technologies with a defined societal benefit. In the place of technology as an avenue for human salvation and the achievement of transcendence, I view technology as a highly negotiated and ultimately political construct. Committed neither to ethical conservatism nor to economic neo-liberalism, I am recommending a democratic process that integrates social and technical change.

And yet, I fail to see in all of this a specifically European regionalism. The science-based development of technology is rooted in a tradition of truth-seeking, criticism and enlightenment, the same tradition that produced the political constitutions not only of the US and Europe. The critical recognition of limits of feasibility, the creation of spaces for political deliberation and agenda-setting, and the expectation of societal benefit from public investment hardly reflect merely regional commitments, but are valued in all parts of the world. The question is only whether these commitments are asserted in public debate and the formulation of sound science policy.

NOTES

1   To be sure, the present reflections do not provide an argument why 'smart environments' belong among the current challenges of scientific and technological development and why 'immortality' or 'brain implants' do not. Here, I can merely enjoin my readers at least to insist on such distinctions. I shall provide additional technical detail in the more extensive versions of this text.

2   Also, James Martin in his opening remarks to the James Martin Institute 2006 World Forum called for science and technology to respond to the big problems that are now facing humankind. To be sure, he also referred in his presentation to cognitive enhancements and the like. But he left open whether these can contribute to the solution of the global warming, water supply problems and so on. These enhancement technologies merely took their place on a generously conceived lists of 'things to come', a list that is not to be counted on as we begin to address the big challenges.

3   Interestingly, transhumanist visions do not normally begin by inviting public agenda-setting, but simply by postulating that the technology and thus the transhumanism will come.

4

Personality Enhancement and Transfer

Bill Bainbridge

We can distinguish three interrelated tasks that lead step-by-step from simple tools assisting humans in their ordinary lives to a radical redefinition of what it means to be human:

1 *personality enhancement*: increasing the cognitive, affective and social functioning and scope of a person through artificial means like information technology;
2 *personality capture*: entering substantial information about a person's mental and emotional functioning into a computer or information system, sufficiently detailed to permit a somewhat realistic simulation; and
3 *personality emulation*: allowing a captured personality to act and perceive in the world, by means of artificial intelligence (AI), robotics, sensors or possible biological embodiment.

The convergence of cognitive and information technologies will promote the convergence of enhancement, capture and emulation. As human beings use ever more sophisticated information systems, their tools become extensions of themselves, their personalities expand into cyberspace and they can begin to transcend location in bodies that exist only briefly in a limited domain of space. Information technology applications must become knowledgeable about the user in order to serve him or her, so systems will perform personality capture incidentally to doing other work. This path leads in small, practical steps from minor human enhancements, to self-transformation, to cyberimmortality (Bainbridge, 2002a, 2002b, 2004a, 2004b and 2005).

MOBILE AND UBIQUITOUS INFORMATION TECHNOLOGY

Henry Kautz, of the computer science department of the University of Washington, is a leader in developing computer systems for *assistive cognition*. The aim is to help people suffering from cognitive disorders like Alzheimer's disease, which requires the computer to become aware of the context surrounding the user, to learn how to interpret the user's behaviour and to offer appropriate help when it appears that the person is failing to reach a goal. One project employs a pocket computer with a global positioning system for users travelling around the city of Seattle, not merely to monitor where the user is but to deduce where he wants to go and guide him if he gets lost (Liao et al, 2005). Another research study employs a glove with radio frequency identification equipment to infer from the movements of a person's hand what activity he is engaged in (Patterson et al, 2005). And a new research project funded by the National Science Foundation (NSF) is comparing sensor methodologies for mapping social interactions among a number of people, to understand the dynamics of their social network and their individual positions in it.[1]

At the University of Michigan, Martha Pollack has been working with several colleagues in developing automatic methods for helping elderly people plan their activities, which requires the computer to learn the habits and preferences of the users (Pollack, 2005). Also at Michigan, Edmund Durfee has been involved in projects to develop computerized planning approaches and methods to coordinate the actions of multiple artificial intelligence agents. Now, with NSF support, Durfee and Pollack have teamed up to develop *socio-cognitive orthotic systems*, designed to help individuals and groups manage their social relations.[2] The elderly have one set of problems with social relations, but young people who are strangers to each other have other problems. At the New Jersey Institute of Technology, Quentin Jones is developing a system for students that will map users in time and space, identifying opportunities to introduce students with similar interests to each other.[3]

Rosalind Picard is a pioneer in the field of *affective computing*, developing artificial intelligence systems that sense and respond appropriately to human emotions, and she has established an affective computing research group in the Media Laboratory of the Massachusetts Institute of Technology (Picard, 1997). Picard envisions a time when an artificial intelligence software agent might be able to help a recovering drug addict by providing constant attention, based on a deep understanding of the user's feelings and challenges.[4] Picard's current NSF research grant supports development of a *social-emotional intelligence prosthetic* that would help people who have difficulty reading the feelings and reactions of other people or expressing themselves competently, who are sometimes (wrongly) diagnosed with Asperger's syndrome or autism spectrum disorder. A portable system that is alert to non-verbal cues would observe the user or people he or she is interacting with, then unobtrusively advise the user.[5]

These are only a few of the research projects that are developing new technologies to assist the disabled, and potentially to assist anyone, in conducting some of the difficult personal business of life. Importantly, these systems would be impossible without artificial intelligence, and the AI must adapt to the individual user. Thus, the intelligent agent in the device becomes a reflection of the user, possessing intimate information about the user's feelings and activities. A somewhat different approach involves systems designed to record everything a person experiences, while moving through the complex real world, either to gather information of practical value to other people or to assemble an archive of the individual's life.

The Experience on Demand Project, led by Howard Wactlar at Carnegie Mellon University in the late 1990s, was funded by the US Defense Advanced Research Projects Agency. This project developed 'tools, techniques and systems allowing people to capture a record of their experiences unobtrusively, and share them in collaborative settings spanning both time and space'.[6] Specific applications included for rescue workers, crisis managers and military scouts, who gain detailed information rapidly in operational settings and must share it with other members of the team. For example, the system would be able to connect a dynamic map of someone's movements with pictures of what the person saw at each point in time and space.

Steve Mann's Personal Imaging Lab at the University of Toronto 'is a computer vision and intelligent image processing research lab focused on the area of personal imaging, mediated reality and wearable computers'.[7] The central idea is that a person would wear a lightweight computer with an optical input/output device over one eye, the *eyetap*. The person would see his or her surroundings perfectly well, but the eyetap could superimpose information on the scene as desired, achieving *augmented reality*, for example the names of people the individual meets. At the same time, the eyetap would be taking in the scene and storing it in detail for analysis, long-term archiving or whatever use the person would want to make of it.

Gordon Bell's MyLifeBits project at Microsoft combines experience capture with many other kinds of data (Gemmell et al, 2004; Cherry, 2005). As the project's website explains:

> *Gordon Bell has captured a lifetime's worth of articles, books, cards, CDs, letters, memos, papers, photos, pictures, presentations, home movies, videotaped lectures and voice recordings and stored them digitally. He is now paperless, and is beginning to capture phone calls, IM transcripts, television and radio.*[8]

Another interesting approach records perceptions and behaviours of people as they play video games, including games designed for educational purposes and games specifically created to record people's reactions to realistic scenarios (Yang et al, 2005).

Already, millions of people spend significant fractions of their lives in virtual worlds where all their actions are already recorded and need only be preserved and interpreted (Bainbridge, 2007). Currently, many researchers are developing methods for handling such data, chiefly to understand the dynamics of social and economic networks, rather than with personality capture in mind, and their techniques will undoubtedly be very useful in achieving the goals described here. For example, a team centred at the University of Illinois is carrying out a million-dollar research project using internet server files of Sony's *EverQuest 2* online game, plus questionnaire responses from 10,000 players.[9]

SELF-IMAGE

My own research has focused on information technology innovations to apply traditional social-psychological methodologies to the new task of *personality capture*. I define this as 'the process of entering substantial information about a person's mental and emotional functioning into a computer or information system, in principle sufficiently detailed to permit a somewhat realistic simulation' (Bainbridge, 2003 and 2006). A pair of questionnaire software modules, *Self* and *Self II*, will illustrate a line of research that is now coming to a conclusion.

The *Self* module is an outgrowth of the *semantic differential* tradition of research in social psychology (Osgood et al, 1957; Heise, 1970; Bainbridge, 1994). The word *semantic* refers to meaning, and the *semantic differential* measures the meanings that different things have for an individual, especially emotional aspects. The word *differential* refers to comparative measurement along scales between words that have opposite meanings (antonyms), such as good–bad, weak–strong and slow–fast. As a tool of personality capture, the semantic differential methodology should employ as many scales as practical, in order to identify all the qualities that define a person, and the *Self* module includes fully 800 pairs of antonyms, presented as 1600 separate measurement scales categorized in 20 groups of 80 words (40 pairs) each.

The respondent is asked two questions about each adjective, for a total of 3200 measurements. One asks how little or much the word describes the respondent, on an 8-point scale from 1 to 8, the 'Little–Much' scale. The other asks how bad or good the quality is, the 'Bad–Good' scale. Semantic differential research has tended to find that the bad–good evaluative dimension is the most significant factor in people's judgements of things. Table 4.1 presents data for one individual, summarized by the 20 groups of qualities.

The statistical correlation between the Good and Much ratings is a measure of the respondent's self-esteem, the extent to which the person believes that the qualities he or she possesses are good. Correlation coefficients can range from –1.00 through 0.00 to +1.00. The rows of Table 4.1 have been arranged in descending order of self-esteem, from 0.92 for the Intellect Group of adjectives down to –0.25

Table 4.1 *Summary of one respondent's self-image*

Label	Description	Self-Esteem Correlation	Mean Difference	
			Good	Much
Intellect	mental activity or inactivity, awareness or unawareness	0.92	3.40	3.28
Quality	value, being of use to others or useless	0.91	4.47	2.90
Organization	the personality's structure, order or disorder	0.86	2.90	2.10
Goal-direction	working consistently for goals or lacking goals	0.84	3.20	2.13
Notional	various poetic dichotomies, in the form of metaphors	0.81	2.58	2.00
Judicial	righteousness or unrighteousness, lawful or unlawful	0.78	2.20	1.67
Kinetic	action or inaction, speed or slowness, movement or immobility	0.76	3.20	1.98
Convention	obeying or violating conventional norms	0.74	3.10	2.92
Harmony	living harmoniously with others or in discord	0.70	1.20	1.20
Likeable	attractiveness or unattractiveness, being liked or disliked	0.65	2.67	1.50
Random	numerous dimensions along which people evaluate good and bad	0.62	3.30	2.53
Miscellaneous	many qualities, not easily classified, and their opposites	0.59	2.30	1.85
Benevolent	being helpful or harmful to other people	0.44	1.55	1.38
Temporal	time, whether past and future or changing and unchanging	0.21	3.67	2.10
Sociability	interaction and communication, facilitating or inhibiting	0.20	3.42	1.95
Ascendant	being better than other people or not better	0.09	2.90	1.80
Physical	the person's physical body, greater or lesser	0.02	2.35	1.33
Feelings	many different opposite emotions, high or low	0.01	2.65	1.60
Dominance	power over others or being overpowered by them	−0.05	4.47	1.23
Erotic	arousing others sexually or not arousing them	−0.25	2.85	2.03

for the Erotic Group. This particular respondent rates himself high on intellect but low – indeed negative – in the erotic dimension.

On average, the respondent rated the 1600 qualities 4.51 on the 1–8 Bad–Good scale, which is almost exactly the 4.50 we would expect if the pairs were exactly balanced, such that the Good word in the pair was as good as the Bad word was bad. The respondent's mean on the Little–Much scale is nearly the same, 4.58, and thus reflects only very weakly the mild tendency many people have to accept descriptions of themselves regardless of their content. The mean ratings across the 20 groups of adjectives would not tell us much, because the positive and negative adjectives in each antonym pair will tend to cancel each other out; more interesting is the average difference between the ratings of the two adjectives, and these are the numbers in the last two columns of the table. A high average difference in the mean ratings of antonyms indicates that the respondent has a clear image of the particular qualities and that the respondent values the positive ones highly. The *Self* software can generate an automatic 'Character Analysis' for each of the 20 groups of antonym pairs, and several other outputs that may guide the individual in self-development.

Self II employs 2000 short phrases that could describe a person, taken from a public-domain collection of psychological items called the International Personality Item Pool, created by Lewis R. Goldberg.[10] Respondents evaluate these on the same two 8-point scales, 'Bad–Good' and 'Little–Much', so the data reflect the individual's values as well as self-image. A major portion of the software produces a 'Personality Analysis' report, giving the respondent's scores on fully 310 multi-item scales organized into 15 personality measurement instruments that Goldberg validated in comparison with standard scales that are unavailable to me because they are covered by copyright. Again, there are two numbers for each scale or item (Bad–Good and Little–Much), so there are 620 composite scale measurements and 4000 individual item measurements. Even with some overlap in meaning across scales, such data provide very detailed information about the individual's personality.

The first hundred items in *Self II* measure the five dimensions of the most influential personality measure, the so-called 'Big 5' model (Wiggins, 1996). There are 20 items for each of the five factors, some phrased positively and some negatively. For example, here are two positive items that measure Intellect:

1   'Can handle a lot of information'; and
2   'Love to think up new ways of doing things'.

And here are two negative items:

1   'Am not interested in abstract ideas'; and
2   'Avoid difficult reading material'.

When items are combined to make composite scales, the scoring is reversed on the negative items, and then all the scores for items contributing to the scale are averaged. Table 4.2 shows these mean scores for the same respondent as in the previous section, to permit comparison, and both analyses reveal the respondent values Intellect most highly and considers himself to be an intellectual.

Another way that both *Self* and *Self II* can give feedback to the user is by grouping together the qualities that the person rated similarly. For example, a person's *pride* consists of the qualities he or she rates highly on both the Good and Much scales; they are excellent qualities and the person possesses them. Similarly, the software can group the items the person thinks are very good but feels he or

Table 4.2 *The respondent's 'big five' dimensions of personality*

Big 5 Dimensions	Average		Rank	
	Much	Good	Much	Good
Intellect	6.60	6.60	1	1
Conscientiousness	4.85	5.55	2	2
Emotional Stability	4.25	5.30	3	3
Agreeableness	3.80	4.30	4	5
Extraversion	3.65	5.25	5	4

she lacks – or the bad qualities he or she possesses – which could be a guide for efforts to change the personality.

The two *Self* modules are among 11 that resulted from my research and development programme. Several of these programs were published in 2002 on a CD enclosed with the anthology *Computing in the Social Sciences and Humanities*, and all 11 were provided online in 2007 at www.cyberev.org (Burton, 2002). As the site explains, 'The purpose of the CyBeRev project is to prevent death by preserving sufficient digital information about a person so that recovery remains possible by foreseeable technology.'

CONCLUSION

As tools for personality capture, the mobile and ubiquitous information technologies have the disadvantage that they collect vast amounts of data that a person may not have perceived or does not remember, and it will be difficult to base emulation on such data sets until we have carried out many empirical studies on what people actually perceive and remember. The much more traditional social-science approaches are more ready to support emulation in the near term. For example, one line of semantic differential research has developed a grammar of emotions, such that a mathematical model would accurately predict how a person reacted to a complex situation based on how the individual reacts to its separate parts (Heise and Weir, 1999). Altogether, the 11 personality capture modules I have created gather 44,000 measurements of the user's attitudes, preferences, beliefs and self-conceptions. Existing computational methods can predict how the person would respond to other questions and to some kinds of real situations, which is one good step towards being to emulate the personality.

However, mobile AI systems that help people in their daily lives will have very significant roles to play in personality capture and emulation. To the extent that an artificial intelligence learns the user's preferences and acts successfully in the real world on behalf of the user, it takes on some of the user's identity. Feedback from the user guides the system to incorporate the user ever more accurately, and accomplishment of the user's goals is an excellent test of the authenticity of the AI surrogate. My own current projects focus on how people remember and evaluate events that happen in their lives – research in the areas of episodic memory and affective computing – and on human behaviour in virtual worlds. We need a wide range of scientific research and engineering development efforts, not only to develop distinct ways of assisting people that incidentally archive their personalities, or to study personalities through focused studies, but also to combine multiple modalities through convergence of methods rooted in separate disciplines. Only thus will we be able to capture, preserve and emulate human personalities in all their natural complexity.

NOTES

1   NSF Award 0433637, 'Creating Dynamic Social Network Models from Sensor Data'; www.nsf.gov/awardsearch/index.jsp.
2   NSF Award 0534280, 'Multi-Agent Plan Management for Socio-Cognitive Orthotics'.
3   NSF Award 0534520, 'Using GeoTemporal Social Matching to Support Community'.
4   Rosalind W. Picard, 'Helping addicts: A scenario from 2021', http://affect.media. mit.edu/pdfs/05.picard-RWJ.pdf.
5   NSF Awards 0555411, 'Social-Emotional Intelligence Prosthetic', and 0705647/ 0705508, 'Social-Emotional Technologies for Autism Spectrum Disorders'.
6   See www.informedia.cs.cmu.edu/eod/index.html.
7   See www.eyetap.org/.
8   See http://research.microsoft.com/barc/mediapresence/MyLifeBits.aspx.
9   NSF Awards 0628036/0628072, 'Instrumenting Behaviors and Attitudes in Virtual Worlds', and 0729505/0729421, 'Virtual Worlds: An Exploratorium for Theorizing and Modeling the Dynamics of Group Behavior'.
10 See http://ipip.ori.org/.

5

On 'Life-Enhancing' Technologies and the Democratic Discourse: A South Asian Perspective

Shiv Visvanathan

Karl Mannheim, in one of his *Essays on the Sociology of Knowledge* (Mannheim, 1952), wrote about two cultural ways of approaching a problem. He observed that when one mentioned the word 'problem' to a European, he treated it as a riddle, an exercise in metaphysics, a question that you unravel again and again, whereas when one suggested the same word to an American, he reached for his toolbox. For him, a problem was something that needed a technological solution. These two perspectives get repeated in the literature. In the debate on life-enhancing technologies, one senses a tension between an America at one end, bristling with technological promise, and a UK/Europe seeded with philosophical doubt. Any third group, with, say, a South Asian or African perspective, needs to add a different view to the debate.

In confronting these debates from a South Asian perspective, we emphasize a search for frameworks. How do we locate these technological documents as imaginations? What do the imaginations of these life-extension technologies say about the link between cosmology, constitutions and the civics of these worlds? Cosmology deals with the relation of man, nature and God, and is basically religious. Constitutions deal with the normative and legal assumptions surrounding these technologies; one looks particularly at notions of social contract implicit in these technologies and their policies. Civics deals with the everydayness of these technologies, usually incorporated in formal explications of innovation chains. Thus the triptych of *cosmology, constitution and civics* provides a developing world framework for approaching these technologies.

Almost predictably, one begins with language. A South Asian confronting these perspectives lists what Raymond Williams called the key terms of the text. Three in particular intrigue in singularly different ways. In examining the relations of word to world, one is concerned with *life, responsibility* and the *Third World*.

We summon as our first witness my perennial hero Ivan Illich. Illich was obsessed with language. For him etymology was an exploration of life worlds and he felt that nothing was as obscene today as the word 'life'. Borrowing from the work of Uwe Pörksen, a Frieburg linguist and medievalist, Illich called life a *plastic word*. A plastic word is not an ordinary word. It is not a vulgar word, a medieval word or something used in ordinary language. A plastic word has been picked up from ordinary language and fed into a scientific laundry. It returns with a new sense of pomposity and connotation. Illich was thinking of words like 'sexuality', 'crisis' and 'information', and he then added 'life' to the list. He was angry that something as precious as life should become a plastic word. He noted that, 'When I use the word *life* today, I could just as well cough or clear my throat or say "shit"!' (Cayley, 1992, p256).

Illich makes three additional observations. He notes that when someone uses the word 'life', he gives it substantive meaning: 'When I speak for a life … I transform the being whom I would call a person into a life.' In the Bible when Christ said 'I am life', the word meant a relationship to Jesus. But today we use the word life for:

> *a zygote, a fertilized egg … we abuse the word for the incarnate god. To turn an attribute created by Jesus to designate himself into an object which you manipulate, for which you feel responsible, which you manage, is to perform the most radical perversion possible.* (Cayley, 1992, p256)

Illich insisted that to use the word 'life' as above in an ethical discourse is to give a semblance of ethics to an unethical context.

Illich remarks that when we shout we are for life, what the word evokes is something ambiguous and imprecise, a sense of 'I ought to be responsible'. Unfortunately this sense of the perversion of the word 'life' extends to the word 'responsibility'.

In his interviews with Bruno Latour, Michael Serres notes that:

> *In dominating our planet, we have become accountable for it. In manipulating death, life, reproduction, the normal and the pathological, we have become responsible for them. … [W]e have resolved the Cartesian question. How can we dominate the world? Will we know how to resolve the next one? How can we dominate our domination? How can we master our mastery?* (Serres, 1998, p172)

In quoting Illich and Serres, what I wish to emphasize is that responsibility cannot be equated to management. Management is instrumental; it is a world of the toolbox. Responsibility demands metaphysics, which is precisely what science policy lacks as a managerial exercise. It opens the toolbox to solve a riddle. In fact, if we look at our innovation chains or the sequence of technological promises made about biotechnology and nanotechnology, one realizes that innovation chains represent banalized millennial movements. Years ago, in Melanesia, millennialism promised refrigerators and air-conditioners and other goods. Today it promises an equivalent cornucopia of technological artefacts. It is this prophecy element in the innovation chain that we need to unravel further.

Reading through policy texts, one inevitably encounters the Third World as a category. As the site of poverty, malnutrition and digital divides, the Third World seems to legitimize every new technology as inspiration, object and an expression of humanitarianism. There is a cynicism and enthusiasm here that we must point out. It was captured best in a comment that Joseph Weizenbaum made in an interview with *Harpers*. Weizenbaum remarked about computers that:

> *there is a temptation to send in computers wherever there is a problem. There's hunger in the Third World. So computerize. The schools are in trouble. So bring in the computers. The introduction of the computers, be it in medicine, education or whatever, is usually to create the impression that generous deficiencies are being corrected, that something is being done. But often its principled effort only serves to push problems further into obscurity – to avoid confrontation with the need for fundamentally critical thinking.* (Oppenheimer, 2003, p39)

The Third World becomes the fundamental object and legitimization for a technological impetus. And what Weizenbaum said about computers extends to life-enhancing technologies.

We began with words, because words become important in any conversation. But for many in South Asia, the debates on the new life-prolonging technologies may not be so much about content as about context. The recent emergence of such convergent technologies creates as it were a site for new debates in democracy. Democracy today needs to be an open work which reinvents itself over a range of sites. One is no longer content with a ritualized democracy reduced to elections and the debate on human rights. Democracy needs open sites like technological projects, scientific controversies and futures issues to reinvent itself. The debates between the US and European views of converging technologies become another such theatre.

There is, however, something limp in the idea of democracy propounded in these policy documents. One does not deny that they are dutiful, but the expert's idea of the public seems to exhaust itself after predictable salutes to information, participation and popularization. The public seems only a listener, a consumer. It seems composed of stakeholders, interest groups and corporations, but the active citizen, interested in his future and science as a part of it, seems rare. The public is not the domain or repository of expertise. It seems barely more than a crowd or a mob susceptible to irrational responses. To guard against this, technological advocates often insist that *not* pursuing a technology, to declare a moratorium on it, is the worst kind of ethics. The new site for technological debate has to have a more active definition of the public as consisting of those not just with interests but with values, with ideas, with knowledge frameworks critical to these decisions and debates. Otherwise it is democracy that becomes vestigial or rudimentary.

From the South Asian archive of debates on science and democracy, let me introduce two terms that may be fruitful in these contexts. The first is the idea of 'cognitive justice' developed to meet the World Bank ideas of participation.

The idea of cognitive justice insists that every person possesses knowledge and that these forms of knowledge must be considered in debates on science and technology. It holds that tribal knowledge, or women's knowledge, or a patient's knowledge cannot be museumized, or archived merely as voice. It must be acknowledged as theory, as something crucial for a democratic formulation of scientific expertise. People's knowledge cannot be devalued as ethnoscience, but must form part of the decision-making process as an agency, as dialogue, as presence and negotiation. Participation is too rudimentary a notion of presence and dialogue.

Once we introduce the idea of cognitive justice, a new kind of democratic theatre becomes possible. A lot of life-prolonging technology seeks to put an end to human disability. It constructs its own expertise of normalcy. It is built up upon the majority view of the non-afflicted. But how does an afflicted group see their own being and the new technologies that are provided to lessen their distress? Are the afflicted only to be seen as patients and victims and never as persons of knowledge about their being? A disability approach to bioethics becomes guaranteed through cognitive justice.

Consider a second idea. One of my favourite philosophers of science, Zia-uddin Sardar, was once discussing the national health system in Britain. He observed it was a travesty of truth and power if the British medical system provided only Western allopathic medicine to Muslim citizens. Ethnicity is not only a question of rights: it is about the availability of alternative medical systems not only as medicines but as philosophies of healing, pain, death and suffering. Within this view, technologies of longevity have to encounter alternative cosmologies of life and death. The ethnic Muslim is not just a patient and a citizen but a philosopher whose framework of choices about death and disability might be different.

Thirdly, nanotechnological questions of life prolongation have linked micro to macro cosmologies. Longevity has to be explored not just as genetic possibility but as an ecological responsibility. How does the longevity of a community rhyme with ideas of hospitality? One has to ask whether longevity is too anthropocentric an idea of man's relation to the Earth and other species.

A second concept that one hopes to offer to this debate is the idea of *tacit constitution*. Social movements in India have argued that modern constitutions are disembedded entities that need to be ecologically anchored. For example, citizenship exists in industrial time and unless a constitution is ecologically located, the rights of the tribal, nomadic and pastoral groups become problematic. The idea of a tacit constitution argues that a constitution floats within a tacit or understated notion of time, metric systems, calendars and regimes of property. This ganglion of assumptions becomes fundamental in deciding one's constitutional chances in modern life. Tacit constitutions help understand the methodologies of assessment and evaluation invoked by new technologies. For example, one can ask whether the notion of progress and obsolescence is part of the civics of the new technologies. The multiplicities of time become fundamental to determining how to create new technologies. The ideas of time even predetermine the modes of assessment.

These new convergent technologies are all futuristic. But how do we look at the future? Is the future only a specified time of preference or is it a multiple emergent domain? Any future democracy is going to be determined by how we think of past and future. Is the past a collection of dead possibilities? The ethics of the future need to consider hypothetical futures and hypothetical pasts. The Jataka Tales (folkloric Buddhist texts) offer ideas of alternative pasts and alternative futures. Here story-telling about technology acquires a different tenor. Ancestors take on a different meaning.

I remember a section in a Philip Dick novel where he asks what would have happened if the Axis rather than Allies had won the war:

> *It felt odd to be imagining not a hypothetical future but a hypothetical past. The more he thought of it, the more this past and the present that ensued from it could have happened this way.* (Carrere, 2005, p62)

My reasons for emphasizing alternative pasts and futures have to deal with concepts of technological forecasting and assessment. It also emphasizes the difference between the future as *play* and the future as *game*. Playfulness transcends seriousness. It is cosmic. Gods play with the universe in a Hindu sense. Play goes beyond instrumental or bounded possibilities, while a game is bounded and instrumental. Most modules of forecasting are game-bound, predictable, professionalized and instrumental. Democracy needs a greater sense of play. We need not scenarios of the future, not game plans, not Delphi technologies, but sheer playfulness. The metaphysics of the two are different. In fact, we need to emphasize that the Third World futures, particularly, must be playful. The future must also be seen as a festival rather than a protestant exercise in solving poverty, disability or death.

Once one is playful, it easier to read the future backwards. As scenario writers, we generally extrapolate linearly into the future. The future becomes a continuation of normal science by other means, a continuation of given categories. But such predictions don't sense discontinuities, breakthroughs, mysteries, miracles or epistemic breaks. We need a prophetic way of looking at the future in a way that ambushes and surprises us. For example, we can't deal with nanotechnology as if it is continuation of current US policy. We need transformative Cassandras who work backwards. Let me explain.

Hannah Arendt (1958), in her *The Human Condition*, talked of the astronomical distance that made the Copernican Revolution possible. She showed how it was only when we visualized a position or perspective outside the Earth that we could think of a heliocentric view. If one needed an astronautical distance to visualize a Copernican Revolution, one needs a chronophilic perspective to visualize the future. One predicts the unanticipated and then works backward in time to create and anticipate this civics.

I would like to get back to relations between cosmos, constitution and civics more systematically. Reading through reports on life-enhancing, one senses a *déjà vu*, even when they are talking about the future. The myths that these documents report are the same overworked ones – Prometheus, Faust, Frankenstein. Such Stakhonovite myths need to be retired. Second, too much of this dialogue takes place without a sense of shadows and doubles or a mirroring or even inversions of situations. For example, the debates on life-enhancement and longevity belong to a similar cluster of discourses on progress, growth, development and perfection. They look linear and morally innocent unless we confront their obverse side in the world of triage, obsolescence, waste, the defeated and the broken.

There is a third problem which I can't visualize fully. In his *Technology as Symptom and Dream*, Robert Romanyshyn showed how the idea of the corpse created as it were the general cycle of science (Romanyshyn, 1989). Romanyshyn shows how the corpse was an invention of the linear perspective. He then plots out a cycle of embodiments of how the corpse as 'body' created transformations which created the embodiments of science.

Romanyshyn shows that the underside of the mechanical body was the witch, madman, monster and hysteric. Each of the latter produced an eruption of the body to match a category above the line. Now that the body has been fragmented into genetic, zygotic, bionic, electronic, cybernetic and ecological hybridities, one must ask what is the relation of these new constructs to the zombie, the cyborg, the android, the micro-robotic swarm. What needs to be worked out is a new paradigm of 'humanness' and its relation to the above creations.

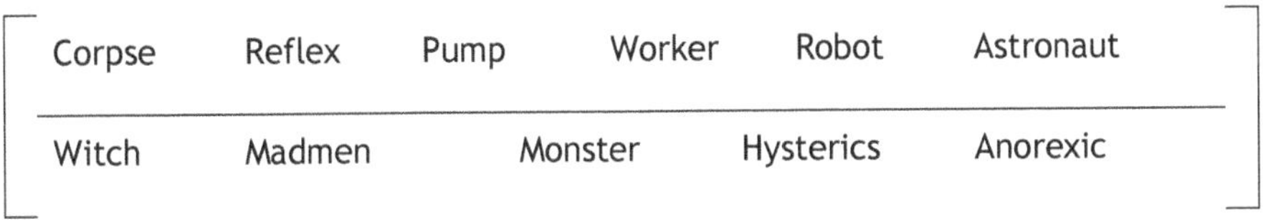

Corpse	Reflex	Pump	Worker	Robot	Astronaut
Witch	Madmen		Monster	Hysterics	Anorexic

Figure 5.1 *Decoding the corpse*

We now need a new parliament of bodies and their relation to what a person is. It is an exercise that needs to be done especially in contrasting the above constructs with what Georgio Agamben dubs that 'HomoSacer', the body as bare life. Otherwise the anthropology of fears and folklores, of science fiction which goes deeper into the recesses of the current scientific imagination, is not available. One could understand evolution better because of the circus, the zoo and the museum. They collected curiosities, deformities and monstrosities. One must ask, 'What is the current equivalent of such systems?'. One needs folklore of the new technologies, some sense of their gossip, their unconscious, otherwise we will be restricted to the aridity of cost–benefit analysis. There is a missing anthropology of life-enhancing technologies waiting to be completed.

There is, finally, a necessity for a reflection on knowledge itself. One is reminded of George Bernard Shaw's comment that England and America are two countries separated by the same language. Today England (and the rest of Europe) and America seem two domains separated by the same technologies. There is convergence, at the level of technology, between nanotechnology, biotechnology, IT and micro-electronics. But European and American frameworks seem to be different. The latter operates with a technological optimism, projecting the idea of the frontier into the new technologies. The former uses the site to introduced ideas of risk and prudence to re-examine the enlightenment dream of science and technology itself. There is a scaling of technologies in terms of the ideas of risk and precaution and complexity, three key terms that provide the gradient for the new life technologies. The idea of cost–benefit seems to cover the normal science of predictability. The idea of uncertainty is covered by the notion of precaution. There is a relation between what is known and what you doubt. But the idea of precaution sees knowledge as a reservoir filling up – it cannot deal with new eruptions of knowledge or with lack of knowledge. South Asian perspectives can add three other dimensions to this sensitivity. The philosophy of the social movements in science could suggest a contouring of risk technologies into domains where cognitive indifference to science, rules of impotence for science, and ideas of taboo where science touches gently can also operate. Linus Pauling once asked for a calculus of suffering for science. What the South Asian perspective seems to imply is that cost–benefit analysis, markets and precautionary principles are inadequate to create this calculus. One might almost say that current economics may be the greatest threat to science. There is a domain where things are not clear. All we have is hunches, life-giving hypotheses, postulates of ecology that need to be played out. Consider the following example.

The Indian Constitution has two sections on rights. The first is a clear-cut chapter on fundamental rights, including rights against torture and rights to property. But the second is a section on 'Directive Principles'. These include non-justifiable rights. They are future dreams, utopias and appeals with a time bar hoping the Indian Government would operationalize them effectively one day. One needs a similar addendum to the regulatory frameworks of the new technologies.

We need a whole generation of thought experiments whereby these technologies acquire an open democratic imagination, where social or cultural imagination of technology gives new assurances against vulnerability, ecocide, uniformity and terror. These new technologies thus become a text, pretext and context for a new vision of science and democracy.

Part III

The Nature of Human Natures

6

Beyond Human Nature

James Hughes

VITALISM AND HUMAN COGNITION

One of the few things that may be unique to *Homo sapiens sapiens* among all animals is our facility for creating abstract concepts, and one of the earliest abstract concepts was probably the idea that there are spirits in the human body, in animals and in things. This hypothesis of the abstract ontological unity and continuity of 'bearness', 'mountainness' or the human soul was a natural extension of human self-awareness, the illusion of a unity and continuity in self-identity (Dennett, 1991; Boyer, 2001). This attribution of an abstract ontology to things has its uses, as it allows us to make predictions about how creatures and things of a kind will behave (Dennett, 1987). The belief that others have an inner self similar to our inner self is the root of empathy. But these vitalist illusions can also trap us into positing identities that do not exist, of making inaccurate predictions and persisting with dysfunctional and limiting beliefs.

Human nature is one such limiting, dysfunctional, illusory and inaccurate belief, the inadequacy of which is revealed in the debates over the moral uses of human enhancement technologies. Take for instance Leon Kass's grounding of his opposition to human enhancement in the existence of Platonic ideal types, including a unitary and inviolate human nature:

> [Creatures] have their given species-specified natures: they are each and all of a given sort. Cockroaches and humans are equally bestowed but differently natured. To turn a man into a cockroach ... would be dehumanizing. To try to turn a man into more than a man might be so as well. ... We need a particular regard and respect for the special gift that is our own given nature. (Kass, 2003)

Without ever clearly defining what this human nature is, Kass deploys the concept to both separate us from our continuity with other animals and bar any improvement in our condition. When exactly does a man's evolution into a cockroach violate his human nature? Is it the loss of a skeleton, the growth of the carapace, the hairy legs or the compound eyes? Can I have tiny antennae, but not big ones? Or is it simply the obsessive compulsive fixation on the scent of food and avoiding light? Likewise, when does man become dehumanized in becoming more than man?

Few proponents of a distinctive and unitary human nature or soul attempt to answer these questions, because they do not have a clear definition of human nature to begin with. They can't specify when hominids got human nature or a soul, or which specific transhuman modifications would rob us of this vitalist essence. They can't agree which aspects of the mind and behaviour are part of the soul or human nature, and which are unnatural, or how parts of human nature might also be shared by other animals. Only after we have deconstructed their illusory theory of a human nature can we begin a serious discussion of the qualities of the human condition worth preserving.

Human Nature Has No Clear Definition

One clear problem with the idea of human nature is that, despite thousands of years of investigation, and intimate access to the subject of investigation, there is no agreement about what human nature is. Are we innately good, compassionate and altruistic, or evil, sinful and selfish? Is moral striving a liberation of our true human nature from sinful influences or capitalism, or is moral behaviour a persistent struggle of the good in human nature against its dark side? Are we a blank slate, morally and behaviourally, or inscribed with all our personality traits, and even beliefs, at birth? Some writers identify human nature with the apparently distinctive human capacities for cognition, language, tool-use, and the creation of meaning and categories, while others include physiology that we share with other species, such as mortality, limbically mediated emotions, and our genetic predispositions for altruism and aggression.

Cognitive neuroscience, ethology and evolutionary psychology are attempting to specify the exact structure and epidemiology of human cognitive traits, and clarify which capacities and impulses are genetically innate and which are plastic or learned. These efforts continue to generate enormous insight, but they have also given some succour to advocates of human nature and natural law, even though these sciences simultaneously challenge the traditional understandings of free will, personal identity and human exceptionalism. Nonetheless, the re-reification of human nature by some evolutionary psychologists has led them to echo Kass's pessimism about human enhancement. Leading evolutionary psychologist Stephen Pinker, for instance, says:

After decades of exile, the concept of human nature is back. It has been rehabilitated both by scientific findings that the mind has a universal, genetically shaped organization, and by philosophical analyses that have dispelled the fear that the concept is morally and politically tainted. So if human nature exists, can it be changed? Attempts to redesign human nature ... are generally recognized as futile, dangerous, and unnecessary to achieve moral and political progress. (Pinker, 2003)

This powerful desire to reconstruct the *Ought* on the genetic *Is* gives us reason to question the claim that evolutionary psychology has revealed a unitary and universal human nature. In *Adapting Minds*, David Buller's (2005) careful deconstruction of evolutionary psychology, the author argues that the field has generated little evidence that human beings have specific genetically driven modules adapted for Pleistocene existence. Rather, Buller believes the distinctive achievement of human evolution was the development of general cognitive plasticity, a dynamic adaptive intelligence which has allowed human beings to invent and reinvent ourselves.

Of course it is true that there are myriad genetic, hormonal and physiological features that shape our desires, thoughts and behaviour, some of which we share with most other human beings. But this constellation of influences fails as a theory of human nature on both analytical and normative grounds. It fails analytically because it posits a vague constellation of species-typical traits which had no clear beginning, are not actually species-specific and are not clearly threatened by any specific enhancement. Normatively the argument fails because we are not morally bound by our genes.

HUMAN NATURE: NO CLEAR BEGINNING AND NO CLEAR BOUNDARY WITH OTHER SPECIES

There is no clear beginning for human nature or the human species. There was, we can assume, no day when all our hominid precursors gave birth to modern humans with opposable thumbs, hidden oestrus, upright posture, language ability, abstract cognition and tool-use. These traits may have emerged abruptly in evolutionary time, but the periods were still tens or hundreds of thousand of years. Which of our grandmothers or grandfathers would the defenders of human nature determine finally had 'it', and were not just savage beasts like their parents? Our branch of the evolutionary tree shows continuous change, right up through the last 15,000 years (Philips, 2006). Without specifying which traits confer membership of humanity, it is not clear whether our genetic differences from Pleistocene humans mean we share human nature with them or not. Did the recently discovered tool-using 'hobbits' of Indonesia, *Homo florensis*, have human nature?

Similarly, we share with primates almost all the qualities that allegedly make us special: self-awareness, culture, language and tool-use. No, they aren't good at abstract reasoning or grammar, but then neither are small children, the demented or the developmentally delayed, and yet they apparently have human nature.

Accepting that the things we value and attribute to human nature are actually shared continuously with non-human ancestors and contemporaneous species is not a devaluation of those traits, or of humanity. In fact it is only by affirming the value of reason, language, compassion and culture-making that we can build an ethical framework to guide human enhancement technologies.

HUMAN NATURE HAS NO CLEAR ENDING

Without a clear definition of human nature, or specification of the things of value, the opponents of human enhancement technology flounder in defining which enhancements cross the line. Francis Fukuyama and the US President's Council on Bioethics see the line being crossed with Ritalin, antidepressants, anti-trauma drugs and preimplantation genetic diagnosis, while others focus further along on the advent of superintelligent immortals and human/animal hybrids. As David Reardon, an anti-abortion activist who is promoting an amendment to the Missouri constitution to forbid human genetic engineering, cloning and transhumanism, has said:

> Any ethic that fails to (1) define human nature and (2) assign some value to protecting human nature inherently lacks the ability to find any limits on the justifications that can be offered to alter or destroy human nature, human beings or humanity. (Reardon, 2005)

Reardon is, of course, completely wrong. Although human nature is being deployed to stop enhancement, the vague and chimerical concept provides no clear lines or policy conclusions. It is only when we let go of the notion of a unitary and inviolate human nature that we can turn to the challenge of delineating which features of embodied human existence are so important that we want to preserve them from technological modification, and which are so central that we want to encourage their enhancement and further evolution.

HUMAN NATURE IS NOT NORMATIVE

Even if we do have some clear set of evolved traits that are distinctively human, they are not normatively binding on us (Bayertz, 2003). To the extent that we are born with impulses for aggression, racism or selfishness, or limits on our capacity for wisdom, awe or compassionate action, we may in fact be morally obliged to modify human nature (Savulescu, 2005).

The boldest and most interesting defence of the naturalistic fallacy of a moral imperative of human nature comes from Francis Fukuyama. Fukuyama argues that human rights and social solidarity are grounded in a shared human nature. Any effort to tinker with human nature will erode social solidarity and lead to totalitarianism. But he explicitly refuses to define human nature, calling it simply 'Factor X':

> Factor X cannot be reduced to the possession of moral choice, or reason, or language, or sociability, or sentience, or emotions, or consciousness, or any other quality that has been put forth as a ground for human dignity. It is all these qualities coming together in a human whole that make up Factor X. (Fukuyama, 2002)

This argument for human nature as an ineffable gestalt is very convenient. If human nature were the sum of these features rather than their irreducible whole, then they might be individually improved, and human nature with them. If human nature was self-awareness, empathy and the ability for abstract thought, for instance, then a green-skinned, four-armed transgenic could still be part of the Jeffersonian polity, and a superior citizen if she was smarter and more empathic.

But Fukuyama's Factor X is also a unique argument that the diversity of humanity must stay within its existing standard deviations from the mean of human traits. This allows Fukuyama to answer the challenge that a normative human nature excludes some existing human beings who don't fit this ideal typical model, such as the disabled. Variation in intelligence, longevity or morphology are OK, so long as we stay within our existing parameters of variation. Although our social unity can apparently still encompass conjoined twins, amputees, people born with fur or tails, and the developmentally delayed, mentally ill and extremely smart, too many kids on Ritalin or too many 130 year olds would apparently break the bell-curved social contract.

But if people four feet tall can feel solidarity with people seven feet tall, why can't the average person be six feet tall instead of five and a half? Why would everyone enjoying the happiness or intelligence experienced by the luckiest 1 per cent of the population fracture humanity into racial subgroups? Certainly, the sudden adoption by a minority of superintelligence, immortality and uploading would challenge existing understandings of shared citizenship, just as shared citizenship had to be forged across racial differences in the past. But human enhancement technologies pose no challenge to Fukuyama's normative standard deviation if all members of a society become more intelligent, long-lived and beautiful, and gradually move the bell curve to the right.

For Fukuyama and the other bioconservatives, this blurring of the line between ur-humans and post-humans is even more horrifying than the emergence of an entirely separate post-human species. Since all good flows from the people of our race having pure Factor X, and race pride in the goodness of our shared Factor X, it must be protected from the complexities of a multiracial society and even more from race-mixing contamination.

THE INESCAPABLE RACISM OF THE HUMAN NATURE CONCEPT

The use of the concept of human nature today is, we see, inescapably racist, *human-racist*, with the same consequences for tyranny, violence and suppression of human diversity as the ideology of European racism before it. The human-racists are more inclusive racists than their forebears, but racists nonetheless in their effort to ground solidarity in biological characteristics instead of shared recognition that another being has self-awareness, feelings and thoughts like our own. We hear in the panicked demands to ban the mixing of human and animal DNA striking echoes of the demands to protect the purity of the white race from mongrelization. The root of this racialist anxiety was laid bare in Mary Douglas's work (1966); it is the taboo on the violation of categories, the ritual taboos against blurring of lines between male and female, white and black, animal and man, and man and the gods.

In fact, Yuval Levin, a staffer on the President's Council on Bioethics, explicitly embraced Douglas's analysis as the mission of 'conservative bioethics' in the inaugural issue of the bioconservative journal *New Atlantis*. The goal of conservative bioethics, he says, is to defend the taboos which:

> stand guard at the border crossings between the realm of the properly human and those of the beasts and the gods. When the boundaries are breached, when degradation or hubris is given expression, our stomachs recoil. (Levin, 2003)

This alleged self-evident repugnance is the same rationale for bans on race-mixing given by all racialists.

The irony is that human-racism is being promoted by some progressives precisely as a means to unify humanity through 'species consciousness', just as white American identity was used to meld together Poles, Irish and Italians, and pan-Arabism and pan-Africanism were promoted to transcend nationalism and tribalism.

The doctrine of a unifying human nature has also become an unquestioned assumption in human rights discourse. For instance, the United Nations *Universal Declaration on the Human Genome and Human Rights* (United Nations General Assembly, 1998) says, 'The human genome underlies the fundamental unity of all members of the human family, as well as the recognition of their inherent dignity and diversity.'

As with bans on miscegenation, human-racists demand bans on human enhancement technologies in order to protect the purity of the human race. President Bush called for a ban on human/animal hybrids in his 2006 State of the Union message, and Missouri has become the first US state to consider a ban on human/animal hybrids, cloning, human-genetic modification and transhumanism. Bioethicists George Annas and Lori Andrews have been working with an international network of opponents of human enhancement towards an international treaty to make human genetic modification a 'crime against humanity'. Genetic enhancement, they say:

> can alter the essence of humanity itself (and thus threaten to change the foundation of human rights) by taking human evolution into our own hands and directing it toward the development of a new species, sometimes termed the 'post-human'. ... Membership in the human species is central to the meaning and enforcement of human rights. (Annas et al, 2002)

Again, like the white supremacists, Annas justifies the suppression of post-humanity on the grounds that post-humans are destined to engage in race war to enslave or exterminate the pure humans:

> The post-human will come to see us (the garden variety human) as an inferior subspecies without human rights to be enslaved or slaughtered pre-emptively. It is this potential for genocide based on genetic difference that I have termed 'genetic genocide' that makes species-altering genetic engineering a potential weapon of mass destruction. (Annas, 2001)

THE VIOLENT POTENTIAL OF THE HUMAN-RACISTS

Is it mere hyperbole to point to the similarity between the race war apocalypticism of the white supremacists and the species-extermination apocalypticism of the bioconservatives? Unfortunately not. Beyond the violence that would be done to human life, longevity and wellbeing by attempts to ban any modification of our chimerical human nature, there is the actual violence that apocalyptic human-racism has already generated, and will generate. Theodore Kaczynski, a.k.a. 'the Unabomber', waged a bombing campaign for *18 years* in the US against scientists engaged in projects that he thought threatened human nature, principally through cybernetics and genetic engineering:

> Human nature has in the past put certain limits on the development of societies. People could be pushed only so far and no farther. But today this may be changing, because modern technology is developing ways of modifying human beings. ... [G]etting rid of industrial society will accomplish a great deal. It will relieve the worst of the pressure on nature so that the scars can begin to heal. It will remove the capacity of

organized society to keep increasing its control over nature (including human nature). (Kaczynski, 1995)

Bombers of abortion clinics are also soldiers in the human-racist effort. In the embryo rights belief system *all* bearers of the human genome have equal moral worth, just as *only* bearers of this human genome have worth. The Christian Right's 'Manifesto on Biotechnology and Human Dignity', which calls for a ban on human genetic modification, makes clear the link they see between defence of the unborn and bans on human enhancement:

> *The uniqueness of human nature is at stake. Human dignity is indivisible: the aged, the sick, the very young, those with genetic diseases – every human being is possessed of an equal dignity; any threat to the dignity of one is a threat to us all ... at every stage of life and in every condition of dependency they are intrinsically valuable and deserving of full moral respect.* (Anderson et al, 2003)

It is therefore no surprise to see common cause being made between pro-choice leftist opponents of human enhancement and anti-abortion activists around their common ideology of human-racism.

BEYOND HUMAN NATURE: THE NEED FOR A (BROAD) NORMATIVE RANGE FOR ACCEPTABLE HUMAN ENHANCEMENT

In conclusion, I am *not* arguing for a laissez-faire approach to human enhancement, unfettered by moral analysis and political regulation. It would be immoral, and perhaps suicidal, for liberal democracies to be indifferent to the directions in which human personalities might evolve using human enhancement technologies. But the concept of a unitary and inviolate human nature is fundamentally the wrong place to start in the analysis of which aspects of human life we want to preserve, suppress or extend. Rather, we need to make clear that it is our capacities for consciousness, feeling, reason, communication, growth and empathy, all of which we share to a greater or lesser extent with other animals, which we are willing to use our technologies and the agencies of our collective suasion – legislation, regulation, social norms and economic incentives – to encourage. It is greed, hatred, ignorance, violence, sickness and death which we wish to discourage, whether part of human nature or not.

Yes, as a part of that project, we must take account of the insights of neuroscience and evolutionary psychology, even if the efforts to mould them into a natural law is wrong-headed and flawed. Understanding the way our genetic constitution shapes our thought and behaviour is essential if we want to use human enhancement technology to improve the human condition, and pursue moral goals that were impossible before human enhancement. So I will close with Peter Singer's closing thought in his essay 'A Darwinian Left' (2000), which argues that the Left must accept that utopian projects have indeed crashed on the shoals of intractable innate human characteristics. But, he says:

> *there may be a prospect for restoring more far-reaching ambitions of change. We do not know to what extent our capacity to reason can, in the long run, take us beyond the conventional Darwinian constraints on the degree of altruism that a society may be able to foster.* (Singer, 2000)

I hope, with universal access to human enhancement technologies, we will soon find out.

7

Are Disabled People Human?

Rachel Hurst

How human beings have evolved as a species is still a mystery, although there are many theories. What we do know is that we have bigger brains than other animals and share much of our DNA. And we have evolved into *Homo sapiens* – a species with physical bodies, brains and spirit, and abilities for social organization, development, empathy and love.

One of our characteristics is the ability to discriminate and to make value judgements. We do this first as newly born babies when we quickly learn to discriminate who is our mother on whom we depend for nurture and survival. We do it later with our wider communities when we ascertain who we want to have as friends. Whether we make these choices for reasons of survival or to verify our own self-worth is debatable. There is no doubt, however, that we discriminate and make value judgements about other human beings all the time. Some judgements are positive, some negative. We discriminate in our actions – separating ourselves from others – or we do it verbally. Whatever the root causes of our negative or positive discrimination, we justify ourselves through formulating myths and stereotypes about the objects thereof.

We make people into icons, celebrities, heroes and heroines. We also denigrate others, both individually and as groups, by defining them and treating them as less than human.

Disabled people have always been seen as different from other human beings. Legislative and social responses to disabled people's needs and rights have separated or isolated them from their communities. In some places and centuries, disabled people have been seen as quasi-gods; in others disability has been seen – and is still seen – as the embodiment of sin. Disabled children have been left to die on Spartan hillsides or at the gates of cities and temples (Fletcher, 1995). Disabled people led the terrible line of men, women and children into the Nazi gas chambers

(Gallagher, 1995). Disabled people have been isolated in institutions, at the back of huts or in inaccessible environments – even isolated in their own homes. Religious and moral teachings say that the sin that disabled people embody can be the instrument through which non-disabled people can achieve greater humanity or sanctity; rehabilitation and community services have built a professional and well-paid hierarchy that has not equalized disabled people's opportunities and has built ever stronger separate and dehumanizing environments.

Legal systems and statutes throughout the world specifically deny justice to disabled people, saying, among other things, that it is all right to kill a disabled child or adult, though it is considered immoral – and illegal – to kill a non-disabled person. Recently in the UK alone, judges have sanctioned the termination of life of two disabled children and have allowed a mother who killed her two disabled sons aged 23 and 20, a man who killed his mentally ill wife, and a father who killed his happy, playful son to go free from court with non-custodial sentences. These stories are just the tip of the iceberg. There now exists an international database of violations of disabled people's rights which has cases affecting over 2 million disabled people. Ten per cent of those violations have resulted in death and 35 per cent in degrading and inhuman treatment (Light, 2004).

Scientific genetic advances pose a further threat to disabled people's humanity, as genetic 'faults' are seen in terms of potentially disabling impairments, and the eradication of these 'faulty' genes seen as the only solution for the advancement of the human race – for the good life and the pursuit of happiness. If we discriminate over the survival or characteristics of the foetus or embryos, then we are seeing them as potential human beings and therefore we need to consider them as human beings, regardless of whether they are legally so. The problem is that impairment and disability are seen as the same thing. Yet in reality, everyone has impairments, either physical or behavioural or both. And we see those impairments as only part of our humanity, not the whole of our personhood. Disability is the social response to our impairments – the responses of attitudinal and systemic discrimination, prejudice, stigma and fear (Barnes, 1991). Disability can only be eradicated through social change and justice and equality for those who are deemed different or deviant because of their impairments, not by fixing impairments.

This subordinate status of disabled people and other groups of human beings and an understanding of their shared inequality and injustice can be measured – and hopefully redressed – by the universal human rights instruments. Society has recognized this in relation to women, ethnic minorities and children, but it is only now that disabled people are being seen as having the same inalienable right to be considered human. As Leandro Despouy, the UN Special Rapporteur on Disability wrote in the introduction of his report to the United Nations in 1993:

> *The treatment given to disabled persons defines the innermost character-*
> *istics of a society and highlights the cultural values that sustain it. It*
> *might appear elementary to point out that persons with disabilities are*

human beings – as human as, and usually even more human than, the rest. (Despouy, 1993)

SURVIVAL OF THE FITTEST

Is this separation of disabled people from the rest of humanity a part of social survival and a natural evolutionary mechanism? Is invalidation and discrimination against others an identifying characteristic of a human being – that is, does deploring difference make you a fully paid-up member of the human race? Or is this separation of disabled people based on a genuine desire to end human suffering, to pursue perfection, to do no harm?

Certainly we want to continue our species, our tribe, our nations. Undoubtedly this survival instinct has ensured that group dynamics have flourished through specific cultural and behavioural guidelines to ensure recognition of what constitutes good breeding/survival characteristics.

Unlike ladybirds, human beings do not use the number of their spots to assess good characteristics, we use fashion, lifestyle and status. We create myths, stereotypes and customs for rejection of the unacceptable and for social elevation of those we see as outstanding. In order to ensure social conformity to these survival guidelines, we introduce behaviour guidance or laws – mechanisms for rejecting or excluding any individual or group which is seen to threaten that survival: immigration laws, apartheid regimes, anti-social behaviour orders and so on. Inevitably these behavioural guidelines are set by those who consider themselves the dominant survivor group and are in control of the social organization, even the channels of democracy (Kallen, 2004).

PURSUIT OF HAPPINESS

Today's society is deeply wedded to the pursuit of happiness, where pain and suffering are seen as abnormal and traumatic and the sufferer is a 'victim', 'vulnerable', in need of relief and counselling. Under earlier conditions of poor maternity care, working conditions and diet, infant mortality and life expectancy were seen as the measures of life quality; now, a good quality of life is assessed by the absence of disability, and both the health and economic status of nations are quantified by disability adjusted life years (Metts, 2001). Following the example of The Netherlands, several countries or states are pursuing or have introduced legislation to allow voluntary euthanasia or assisted suicide. But it is hard to assess whether this is being done to cut the costs of palliative care in spiralling health budgets or to respond to the fears of those who have seen their loved ones die in unacceptable conditions due to the absence of proper palliative care. What is clear is that people no longer see pain, suffering and death as part of the normal human process, events from which they can develop a greater understanding of life and their own personhood, but as events which they want to control and then have the power to reject.

When analysing why disabled people are seen as non-human, we cannot avoid the classic debate of a human being's intrinsic worth as opposed to instrumental moral value. Should life be valued in itself and protected under a set of moral codes, or is it only valued in relation to others? Undoubtedly there are many who believe that life is only of positive value in relation to a quality of life that they themselves think necessary for a 'good life' and in relation to others who share those values.

DEVIANCE

Sagarin (1971) was one of the first social scientists to argue that negative, invalidating judgements that prevailed with regard to racial and ethnic minorities were also applicable to similarly disadvantaged and stigmatized non-ethnic groups such as women, children, the aged, homosexuals, alcoholics and disabled people. Traditionally, these non-ethnic groups were analysed by social scientists as 'deviant' and anti-social. This medical model located the deviance in the individual's pathology and assumed that the identifiable impairments, abnormalities or disorders in the individual were the determining cause of deviance. This traditional view has been vehemently contradicted since the 1960s (Becker, 1963), however, as deviance has been analysed within a social construct and the recognition has arisen, both in the groups concerned and the wider public, that the negative stereotype of deviance has further discriminated and invalidated these groups from the dominant group.

Whilst this social construct and negative stereotyping has been readily accepted and understood in relation to most 'deviant' groups, disabled people have continuously, despite substantial evidence to the contrary (Oliver, 1990), been viewed in relation to the medical model. As was outlined in the introduction, this has been a major factor in the dehumanizing of disabled people and led to geneticists' and bioethicists' concentration on disabling impairment as deviance.

EUGENICS

Francis Galton's theory of eugenics was to better the human stock of the nation. He wanted to improve good genes, rather than eliminate bad ones, but the elimination of deviance, depravity, destitution, vagrancy and criminality was widely supported by the leading thinkers of the day who were members of eugenic societies in Europe and North America. They saw this elimination, as well as the promotion of health and wellbeing, as a moral and economic imperative for their countries. This support of eugenics led to the compulsory sterilization of 'degenerates', who were mostly disabled people, in the UK, US and Scandinavia and during the Third Reich.

To achieve 'goodness in birth' (the Greek translation of eugenics) and to relieve pain and suffering is the laudable aim of most geneticists and bioethicists. It is certainly the objective of gene therapy. But it does appear that the baby is going out with the bath water – literally. The impaired gene/chromosome/embryo/foetus is not seen as part of a potential human being, but as a flawed structure to be changed or eradicated and so surrounded by negative prejudice that to eliminate it seems to be the morally valid and best response for the individual, the family and society as a whole.

DIFFERENCE

Is impairment or difference a bad thing? Is a non-sentient human being no longer a human being? Or is someone who looks different, operates differently or needs a different environment intrinsically bad? How can we assess quality of life. Is one person's heaven another person's hell? What is perfection in a human being – is it just physical and intellectual abilities and wealth or is it moral worth? When we learn that the greatest number of suicides is among those with high IQs (Persuad, 2006), it becomes clear that intelligence does not always go hand in hand with happiness.

Who could assess the quality of life of David Glass, a 14-year-old boy with hydrocephalus? The doctors made no attempts to help him live until 50 minutes after his birth, when he began to cry. Then 12 years later, when hospitalized with a chest infection, the doctors tried to ensure 'a death with dignity' by injecting him with diamorphine. But as his mother says:

> David gives me love and honesty. When he smiles, I know it's not because he wants a bag of sweets, it's because he loves me. He doesn't have an evil thought. He isn't driven by power or money. He is just a genuine, nice kid. (Levin, 2000)

Luckily, his family fought – literally – to keep him alive and for the drip to be withdrawn. For the two years since that incident, he has played his role within a loving family. David is making as good a contribution to life as any other son or daughter – it just happens to be different.

The level or degree of impairment is also seen as a reason for elimination – though not for cure. Again, what constitutes severity is assessed by prejudicial notions of what constitutes pain and suffering or a poor quality of life. UK abortion laws, for instance, allow for abortion of a viable foetus, if 'there is a substantial risk that if the child were born it would suffer from such physical or mental abnormalities as to be seriously handicapped' (UK Abortion Act 1967). On

16 March 2005, the Crown Prosecution Service concluded that they would not bring a prosecution against two doctors who undertook an abortion on a 28-week unborn baby with a cleft palate, despite a cleft palate being non-life-threatening and remediable.

Certainly arguments could be made that society cannot economically afford to keep such people alive – and China felt that so strongly about girl babies that they limited families to one child within a culture of preference for a male child. That policy dangerously affected the gender balance and, coincidentally, did nothing for the economy. China found, as had most of those engaged in social development, that to improve the economies of any nation, it is imperative that women are supported and maintained in their equal and, quite often, dominant, role in community and family activity (United Nations, 1995). And as the past President of the World Bank, John Wolfensohn, said:

> Unless disabled people are brought into the development mainstream, it will be impossible to cut poverty in half by 2015 ... goals agreed to by more than 180 world leaders at the UN Millennium Summit in September 2000. (Wolfensohn, 2002)

If we say that disabled children should not be allowed to live because of their cost to the health service, then are we going to say that anyone who is badly wounded in a war should be left to die. What of those people who are severely hurt in a train crash or road accident? If cost becomes a factor in the selection of who is a fully contributing human being, then rights and democracy fly out the window and biotechnology becomes a form of social control.

In today's rights-based world, nobody would publicly say that to be different in relation to gender or ethnicity was a bad thing. But it is an accepted belief that to be disabled is bad and disabled people should be relieved of their suffering and this 'doing good' gives humanitarian gold stars to those who, in effect, seek elimination of these negative traits. The positive attributes of disability seem to be too much for many people to believe or understand.

Public statements regarding the badness of disability, however outraged disabled people may be by them, are common. In contrast, Christopher Newell, a leading disability activist in Australia, said at an exhibition on the genetic revolution:

> It is one thing to have treatment for a medical condition, but another when genetic makeup is such a compelling tragedy as to remove the claim to life itself. And in screening out disability, are we also screening out quintessential human qualities such as courage, persistence and fortitude? Are we valuing technology above the inherent dignity of human life? Could biotechnology ultimately signal the end of genetic diversity? (Head, 2006)

Both Newell and Despouy, quoted at the beginning, are saying the same thing: genetic diversity is necessary for creating a vibrant and sustainable society, and disabled people, regardless of the level of their impairment, have essential gifts to bring to human life.

As disabled people throughout Europe have said in their statement about bioethics:

> *Many disabled people are only alive today because of scientific progress generally and new medical techniques in particular, so of course we wish to promote and sustain such advances where these lead to benefits for everyone. But we want to see research directed at improving the quality of our lives, not denying us the opportunity to live.* (DPI Europe, 2000)

8

Biotechnology and its Spiritual Opposition

Lee M. Silver

THE AGRICULTURAL REVOLUTION AND HUMAN CIVILIZATION

The long history, legacy and pervasive impact of biotechnology on humankind and the biosphere as a whole is not fully appreciated even by most well-educated people. Although it is commonly thought to be an invention of 20th- and 21st-century scientists working inside brightly lit high-tech labs, biotechnology was first developed at the end of the last ice age, between 8000 and 12,000 years ago, at multiple independent locations around the world (Smith, 2001; Fedoroff, 2003).

As the Ice Age drew to a close, human populations grew rapidly and spread out across the subtropical and temperate zones of the Americas and Eurasia. Hundreds of species of large game animals were hunted into extinction, and edible vegetation was over-foraged. In the past – when the human footprint on the world was smaller – a tribe could simply get up and move from a nutritionally depleted habitat onto virgin land where natural resources were still plentiful. But now, habitable virgin lands were exhausted as the entire biosphere was pushed beyond its human carrying capacity (Fagan, 1996). Any other species would have collapsed under the weight of its own voracious appetite, but human genes had endowed human beings with the capacity to initiate a revolutionary lifestyle change that blew apart the traditional equations of adaptation and survival. Instead of fitting into a natural world as best as they could – like every creature before – the human species *consciously* took control away from Mother Nature and into its own hands through a process we now refer to as the agricultural revolution.

The agricultural revolution emerged out of the human discovery of genes – the invisible abstractions that carry specific characteristics of plants and animals from one generation to the next. Genetic conceptualization allowed people to create novel organisms expressing *domesticated* characteristics built to satisfy both human needs and their newly emerging desires. In central America, a slender weed named *teosinte* with a few dispersible hard seeds was transformed into cobs of corn with tightly attached kernels that only come off when we went them to (Fedoroff, 2003). Corn looks nothing like the *teosinte* weed it was engineered from. In fact, scientists wouldn't even know they were related without the tools of modern genetic analysis.

The independent discovery of genes allowed people living in the Middle East to transform an entire series of weeds into wheat, peas, chickpeas and lentils (Smith, 2001). In South America, shrubs from the poisonous nightshade family (*Solanaceae*), with tuberous roots, spiny branches and bitter berry-sized fruit, were transformed into juicy red tomatoes, potatoes, sweet potatoes and peppers. In South Asia, the chromosomes of wild inedible weeds from Malaysia and India were combined to create banana trees so overloaded with DNA that their fruit can't produce any seeds and are completely sterile. Propagation of seedless bananas over the subsequent millennia has depended on the human application of cloning. And in Central Asia, grafting technology was perfected for growing bi-species trees with roots from hearty but inedible crab-apple stocks and stems from mutant crab-apple trees bred to produce what we've come to know as apples.

Like plants, animals were modified by people to create bio-factories that are ever more efficient in the generation of food and other valued products. But in addition, animals presented people with opportunities to control another set of characteristics – behavioural instincts. Indeed, animal domestication is defined by genetic modification of behaviours. Wild, human-threatening and human-fearful instincts are always eliminated and replaced by tameness, an acceptance or desire to be near humans, and, often, other specific human-serving personalities.

In the Middle East, oxen were transformed into docile ten-gallon-a-day milk-producing factories with absurdly large udders; their mammary glands were expanded, the amino acid composition of their milk was altered to better suit human nutrition, and the weaning age of calves was reduced so that humans could use them more quickly to convert grass into milk (Beja-Pereira et al, 2003). Also in the Middle East, a hair-covered goat was bred into domestic sheep that grow unnatural billowing coats of wool, and wild boars were transformed into pigs that subsist on garbage, breed profusely, mature rapidly and live comfortably in close association with people. As a result, pigs provide more meat for worldwide human consumption than any other animal (Kijas and Andersson, 2001).

Once an animal was domesticated for one purpose, it made sense to select simultaneously for additional products (although modern era farmers have reverted to selecting specialized varieties for one use only). Chickens provide eggs, meat and feathers for insulation; cows are used for meat, milk and leather; and sheep give us meat and wool. Finally, after the domestication of plants and animals, ancient biotechnologists discovered and developed the biochemical capacity of the third 'kingdom' of living organisms – microbes – to create new products like cheese, vinegar, wine and soy sauce that satisfied the need to prevent food spoilage, as well as the human desire for novel gustatory and drug-induced mood experiences.

Biotechnology's significance is hard to exaggerate. Its invention represented a fundamental turning-point in the story of the human species and provided the gateway to civilization. With the domestication of plants and animals, tribes were no longer limited in size by the nutritional capacity of their natural environment. When an undisturbed plot of woods or brush was converted into a cornfield or rice paddy, the yield of edible vegetation could be multiplied by millions. And if properly managed, the same rice paddy could be regenerated year after year to produce the same high yield. Farmers produced more food than was required for themselves and their families, and the surplus became available for trading to specialized craftsmen for other desired objects or services. Domesticated organisms themselves – cows, pigs, chickens, corn, wheat, rice and a few dozen other useful biotech creations – flowed along migration and trade routes across the Americas, Europe, Asia and Africa (Diamond and Bellwood, 2003). Trades became professions that exploded in diversity as tribal settlements grew into villages, villages grew into towns and cities, and cities joined together to become nations with ever more diversified economies and complex technologies (Braidwood, 1952). And although no one knew it at the time, within a brief moment of the history of life on Earth, the relationship of our species to all others was forever changed.

THE GREEN REVOLUTION AND AVOIDANCE OF FAMINE

Throughout the first 10,000 years of the biotech era, the rate of innovation was obviously stupendous, but it was still held in check by the very low rate at which genetic alterations – mutations – arise spontaneously. In the first half of the 20th century, however, a huge technological advance occurred when scientists discovered, by chance, that high energy radiation and certain mutagenic chemicals could induce mutations at a hundred-fold greater rate. By 1967, artificial mutagenic methods had been clearly validated, and an international community of agricultural scientists, representatives of developed countries, and humanitarian aid groups came together to exploit its power for the purpose of improving the lives of subsistence farmers in the least developed countries of the world. The International Rice Research Institute (IRRI) was set up in the Philippines, and the International Maize and Wheat Improvement Center (CIMMYT) was set up in Mexico. Together, these institutes focused their efforts on the three most important crops in the world.

Just as the globally supported effort to develop new crop varieties was getting underway, in 1968 the Stanford ecologist Paul Ehrlich published his best-selling book *The Population Bomb* (Ehrlich 1968). In his opening paragraph, Ehrlich wrote grimly:

> *The battle to feed all of humanity is over. In the 1970s the world will undergo famines – hundreds of millions of people (including Americans) are going to starve to death.*

Thirty-seven years later and doomsday has yet to arrive. Why? Ehrlich's main mistake was spiritual rather than scientific. In his heart, he couldn't accept the idea that biotechnology might benefit humanity. What actually happened to food production during the three decades after *The Population Bomb* was released is now called the Green Revolution.

Beneficent, publicly supported Green Revolution biotechnology led to the creation of thousands of varieties of crops with increased disease and pest resistance, increased tolerance to drought and poor soil conditions, and increased nutritional value (Khush, 2001). In one striking example out of many, new varieties of rice (previously limited to one crop per year) could now be planted and harvested multiple times each year. Yields doubled and costs of production were slashed as genetically improved seeds were distributed to poor farmers in Indonesia, India, Mexico, and other countries in Asia and Latin America.

Indeed today, anti-biotech activists knowingly take a stance opposite to the one preached by Ehrlich when they argue that 'we can already grow enough food for everyone – starvation is due mostly to the unequal distribution of food, political posturing and the economic power of the wealthy' (Ross, 2001). Indeed they are correct. Current farmland could produce enough food to feed everyone in the world, but that's only because of biotechnological innovation, a fact that biotech opponents typically ignore. They also refuse to understand the fact that biotechnology has never been *just* about making sure there's enough food to eat. When the costs of producing and consuming food are reduced, more money is available for people to spend elsewhere, which allows them to increase the *quality* of their lives.

MODERN BIOTECHNOLOGY

Until the 1970s, changes in the DNA of an organism could not be controlled by conscious beings. But then a new chapter of biotechnology burst onto the scene, as molecular biologists developed increasingly sophisticated methods for precise control over the design and implementation of particular DNA alterations. At first, they learned simply how to move genes from one organism to another. As a poignant example of the power of this first generation, targeted gene-modification technology, a microbe was created that produces human insulin, which provides diabetics with a cheaper, more natural alternative to the pig pancreas insulin they previously required to live a normal life. Microbes that produce many other therapeutic human proteins have also been created in a similar manner by splicing functional human genes into the microbial genomes.

The methods of genetic engineering are now much more sophisticated, allowing the implementation of far more subtle changes than whole gene swapping. The subtlest change of all is the switching of single letters in a sentence of gene code in a predetermined manner, for example from GC**G**AGAGTTC to GC**A**AGAGTTC. Most spontaneous mutations that occur in nature also cause single-letter changes. With intense commercial interest in both pig and cow farming, the new tools of molecular analysis will surely be used to identify many single-letter and multiple-letter changes in the pig or cow genomes that could increase an animal's value, provide a healthier product for humans to consume or reduce the animal's negative impact on the environment. With directed genetic engineering methods, each imagined change could be efficiently implemented in the production of animals with appropriately modified genomes.

In less than 30 years, the power and accomplishments of modern biotechnology have already been mind-boggling. In the agricultural domain, crops and animals have been modified to provide nutrition and calories with enormously improved efficiency at every stage of production, using processes that are much more friendly to the environment than traditional agriculture (Karatzas, 2003; Smirnoff, 2003). Plants and animals are also being deployed as pharmaceutical factories, and large animals such as cows, sheep and pigs are being engineered to produce humanized blood for transfusions, humanized milk to replace infant formula and humanized organs for transplantation (Kuroiwa et al, 2002; Brophy et al, 2003). Not all applications of biotechnology to plants and animals will have benefits that outweigh costs, but each potential idea and targeted implementation can be evaluated on a case-by-case basis, which is more than is required currently for crops derived from randomly mutagenized seeds or new animal breeds. But, ironically, as molecular biology brings precision and transparency to the actual genetic and cellular modifications that biotechnologists perform, it also shines a brighter spotlight on contended connections between organismal life and spirituality.

A WESTERN BACKLASH AND AN ASIAN OPPORTUNITY

The willingness of traditional Christian believers to accept the potential benefits of genetically modified crops finds roots in the Bible. In particular, only man is said to be created in the image of God. All other living things are put on Earth for man's benefit, and man is specifically given 'dominion' over them. So while plants and animals may have been God's creations originally, God has since delegated responsibility for their upkeep to man. In this context, GM crops are not viewed as inherently good or bad. They can be evaluated, instead, based on a rational assessment of costs and benefits. Of the ten countries with the largest areas of GM crops under cultivation in 2004, six were in the western hemisphere (James, 2004). Outside of North America, the four countries in this class – Argentina, Brazil, Paraguay and Uruguay – are all dominated by powerful Catholic hierarchies. In addition, the Asian countries of China and India are ranked at numbers 5 and 7 respectively. But notably absent from the top ten GM producers is any country from Europe. The European rejection of GM crops, I argue, is a consequence of Europe's Christian roots combined with its current rejection of traditional Christian beliefs.

The results of a 1998 survey of religious beliefs in European countries indicates how far Europeans have moved away from their Christian traditions (The International Social Survey Program, 1998). In Sweden, Switzerland, Norway, Germany, The Netherlands, France and the UK, less than half the population holds a strong belief in a traditional Christian version of God. By comparison, a strong God belief is professed by 78 per cent of Americans and 91 per cent of Chileans (the only South American population covered in this survey). Although the level of atheism and agnosticism in Europe is relatively high, another quarter or more of the population of many Western European countries (Sweden, Switzerland, Austria, Norway and western Germany) 'don't believe in a personal God', but 'do believe in a Higher Power of some kind'.

It appears that many Western Europeans are in desperate need of a substitute to fill the spiritual void left behind when the God of the Bible is rejected. But Western culture is permeated by Judaeo-Christian monotheism, and so the substitution is made most easily through a transformation of traditional Christian beliefs into a post-Christian religiosity. The sacredness of the material human body, symbolized in Jesus, morphs into the sacredness of a material Mother Nature. God's master plan for humanity becomes Mother Nature's master plan for the whole-Earth biosphere. Earth's creatures are now viewed as component parts of Mother Nature's body; if we engineer her genes – the modern analogue of a singular higher spirit – we are liable to upset the natural order.

An example of post-Christian European spirituality is etched into the Constitution of Switzerland through an amendment demanding respect for *l'intégrité des organismes vivants*, the integrity of living organisms, and *Würde der Kreatur*, the dignity of creatures (actually *living nature as a whole*). The amendment was not just a call to alleviate animal suffering or to prevent animal breeding for food (since most who voted for the referendum eat meat). Instead, a majority of the Swiss people felt that their picture-perfect valleys of well-tended meadows, neat farms and grazing cows represented a *natural order* that had to be preserved at a deeper spiritual level. It doesn't matter that every component of this picture is a direct result of human intervention into a previous natural order which has long since disappeared.

The spiritual traditions of the East are diverse, but they all share certain Hindu/Buddhist-derived foundations in contradistinction from Western monotheism. There may be many gods or no gods (depending on the semantic distinction between gods and divine spirits), but there is no Master Creator in the East, nor is there any master plan that we can violate. If no master – or master plan – of the universe exists, the injunction to not 'play God' has no meaning or suasion. Furthermore, in the Eastern worldview, bodies are discarded when they wear out, and their spiritual inhabitants reincarnate within new material beings. In this context, while biotechnology may affect the material, it can't touch the spirits, every one of which has existed from the beginning of time, and will go on existing forever no matter what we do. Consequently, biotechnology is neither feared nor despised by common people in Asia, and it can be pursued without the grassroots opposition typical in Europe.

India will focus primarily on GM crops, while China is cleverly leveraging all biotech fields from embryo research to GM crops to human gene therapy technologies. They have made great efforts to bring home many expatriate scientists from the US and Europe. The Chinese claimed to have an initial concern about GM crops, but most likely, that was a tactic to prevent American companies from entering their market. For economic and political reasons, they have focused on the genetic engineering of rice, knowing that this crop is not a European staple and that other Asian countries are more spiritually accepting.

Generally, non-Christian Asians feel comfortable with all forms of biotechnology, and their governments are poised to leap ahead of Western countries in research and development of plant, animal and embryo engineering. The economic ramifications of cultural differences in spirituality may be enormous: a future in which the West dithers as Asia becomes dominant in both the science and commercialization of biological processes.

Part IV

Longer?

9

Understanding Global Ageing

Sarah Harper

In the latter half of the 20th century, the more developed countries of the world experienced population ageing to a degree unprecedented in demographic history. In the first 50 years of this new century, less developed and transitional countries are predicted to go through the same transition. Such population ageing is arising from a steady fall in both fertility and mortality across the globe, with the exception of sub-Saharan Africa. Europe reached 'maturity' at the turn of the millennium, with more older people (60+) than younger (under 15). It is predicted that Asia will become mature by 2040, and the Americas shortly afterwards. Such global ageing is not occurring in isolation – it is emerging in the context of globalization itself, in a world increasingly dominated by the flow of human and economic capital across national boundaries. Indeed, a key stimulus to such capital flows is the emerging demographic imbalances arising from the differential movement of regions into maturity. Thus while an understanding of the dynamics of globalization is essential to address the challenges and opportunities of ageing societies, so it is also necessary to understand the dynamics of global ageing as a component of globalization.

EMERGENCE OF DEMOGRAPHICALLY MATURE SOCIETIES

By 2000 there were more people aged 60+ than under 15 in the EU 15, by 2040 there will be more older than younger people in Asia, and it is predicted that by 2050 the world's older people will outnumber the young. While in percentage terms, the world's population aged 60+ had risen only from 8 per cent in 1950 to 10 per cent by 2000, this accounted for an increase in numbers from 200 million to 600 million older adults. It is expected that by 2050 the absolute figure will reach 2 billion, another tripling in just over 50 years, when the proportion aged

60+ will have reached more than a fifth of the total global population (21 per cent). The numbers of those aged 80 and above will show an even greater increase, rising from 70 million to 401 million by 2050.

While much of the debate has focused on the ageing of Europe, it is in fact the Asia-Pacific region, currently with 600 million older people, that is the most rapidly ageing world region. It will have some 20 per cent of its projected population aged 60+ by 2050, accounting for two-thirds of the world's 2 billion elders. Of key importance is the speed at which this transition is occurring, with the less developed and transitional countries facing extreme rapidity of ageing. While it took Europe (EU 15) some 120 years to go from a young to a mature population, maturity being achieved in 2000, such a shift in the proportion of young and old will have occurred in Asia in less than 25 years. France, for example, took over 100 years to move from 7 per cent to 14 per cent of its population aged 60+, Japan just 30. While the predicted increase by 2025 in the percentage of people aged 60+ for the EU 15 is around 33 per cent, it is 400 per cent for Indonesia, 350 per cent for Thailand, Kenya and Mexico, 280 per cent for Zimbabwe, and up to 250 per cent for India, China and Brazil. This rapidity of demographic ageing will be one of the greatest policy and institutional challenges for less developed and transitional economies.

Regarding the proportion of the population aged 60+ in the world's oldest countries at the turn of the 20th century, with the exception of Japan, the top 20 are all European. Globally, Italy has the highest proportion of persons aged 60+, primarily a consequence of its low fertility levels. In the EU 15, Italy has the highest proportion of older people (24.2 per cent), while Ireland has the lowest (15.1 per cent). Australia, Canada and the US are all lower, at between 12 and 13 per cent. Interestingly, even the former Eastern European countries have higher percentages than Canada, Australia and the US, but these proportions represent very different numbers of older people. The largest population of older people in the developed world is in the US, with 45 million people aged 60+, followed by Japan with nearly 30 million and Germany with around 19 million.

A society's median age, that is, the age that divides the population into numerically equal parts of younger and older people, provides another measure. All the countries in the developed world have median ages greater than 32. Median ages, however, will increase markedly in some countries over the next quarter-century. Italy, Brazil, China, Mexico and Thailand, for example, will all experience more than a 10-year increase in median ages. Italy is currently predicted to have the highest median age at 52. Japan will reach 50, with most other developed countries, and some Asian ones, attaining median ages greater than 40. Singapore's population structure, for example, has been changing since the 1980s, with a steady decline in the proportion of children and an increase in the proportion of older adults. As a result, the median age of the country's residents has increased by more than ten years over the past quarter-century from 24.4 in 1980 to over 35.5 today. Both Hong Kong and Korea are now over 35.

OLDEST OLD

The growth rate of those aged 80+ is of equal significance. This is the fastest growing age group in the world, with an annual growth rate of 3.8 per cent. Low fertility around the time of World War I and declining mortality rates among this cohort partly explain this: people reaching 80 in the mid-1990s were part of a relatively small birth cohort. Fertility increased again in the post-World War I period, so that at the turn of the century a much larger birth cohort was reaching 80. In just four years, the growth rate of the world's 80+ group thus increased from 1.3 per cent to 3.5 per cent. The projected annual growth rate of this age group is 3.9 per cent until 2010, remaining above 3 per cent until at least 2020. It is predicted that by 2050, 20 per cent of persons aged 60+ will be in this group. Currently 40 per cent of those aged 80+ live in Asia, some 16 per cent in China alone, partly a reflection of China's very large proportion of the total world population; 30 per cent are in Europe and 13 per cent in the US. Japan is predicted to have 40 per cent of its older population aged 80+ by 2030. In recognition of this increase in the oldest old, the UN Population Division is now producing population projections with a final age category of 100+.

It is predicted that by 2050, the population pyramid for the developed world will settle as population parallel lines, with around 10 per cent of the population in each age decade between birth and 100; that of transitional and developing countries will also straighten considerably. Over the next 40 years, however, we shall continue to see a top-heavy pyramid, with a large bulge of mature and then older adults moving up as the dominant population. In the developed countries this is due to the baby boom cohorts of the middle of the last century. In the less developed and transitional countries, on the other hand, this is due to the 'shelf' generation: the current cohort of young reproductive women who, while themselves typically part of large horizontal families with five to eight siblings, have chosen to bear only one, two or even no children.

THE DRIVERS

Global population ageing has been fuelled by a fall in total fertility rates (TFRs), the average number of children that would be born to a woman over her lifetime if she were to experience the exact current age-specific fertility rate throughout her reproductive life. Alongside the well-recognized low fertility of Western Europe, with all countries below replacement level, and southern Mediterranean countries in particular at 1.2–1.4, we see a similar pattern emerging in Asia. Singapore and South Korea have now fallen to below 1.2, while Hong Kong, at just below 1, now has the lowest TRF in the world. The most striking feature of life expectancy rates at birth is not only that everyone born in the developed world, with the exception of the former Eastern European countries, can now expect to live for more than

75 years, but also the high life expectancies at birth in much of Asia and Latin America. Indeed, with the exception of Africa, many less developed countries now have life expectancies at birth of 70 years or older. Furthermore, continued declines in mortality in both the more developed and transitional and developing regions are expected to extend life expectancy at birth to 82 and 75 years respectively by 2050, thereby reducing the gap between these regions. Of particular interest is the rise in healthy active life expectancy, with current predictions for Europe and the US forecasting that both men and women in their early 70s can expect to live well into their 80s, enjoying most of those years disability-free (Manton et al, 2006).

THE GLOBALIZATION OF POPULATION AGEING

As indicated earlier, global ageing is emerging in the context of globalization, and is in itself stimulating the flow of human and economic capital across national boundaries as a result of the emerging demographic imbalances arising from the differential movement of regions into maturity. It is thus necessary to understand the dynamics of global ageing, as a component of globalization, addressing it at the global-institutional, societal-institutional and individual levels. Let us start with the latter two.

At the societal level, demographic change will clearly have significant implications for labour supply, family and household structure, health and welfare service demand, patterns of saving and consumption, provision of housing and transport, leisure and community behaviour, and networks and social interaction. However, as governments and policymakers have awakened to the implications of population ageing, so the 'demographic burden' hypothesis has spread. National health services, and even economies, are predicted to collapse under the strain of health and pension demand, and families will no longer be there to compensate for failing public provision. Above all, ageing is seen as a challenge for the West alone, having little relevance for less developed and transitional countries, and one that can be compensated for by immigration from the young South. The reality, however, is far more complex, and highly susceptible to policy changes. Indeed, understanding the reality of the demographic issues is vital, both for individuals who need to reassess their life courses in the light of the new longevity probabilities, and for governments charged with planning and developing appropriate policy frameworks to address the forthcoming demographic changes, challenges and opportunities. The major concerns are public spending on pensions, high dependency ratios between workers and non-workers, increases in healthcare costs, declining availability of family-based care, and a slowdown in consumption due to an increase in older people and a decrease in younger people. These are dynamics of current cohorts and current behaviour; they are not fixed. Furthermore they are all phenomena which can be addressed by policy, given the political and economic will.

In addition, as already indicated, population ageing is of major concern to countries in Asia and Latin America, where the demographic challenges due to the rapid acceleration of ageing will confront governments already overstretched in their resources by acute poverty and famine, with high levels of infant and maternal mortality, acute infectious diseases, and lack of access to basic sanitation and fresh water. European societies addressed these challenges over 150 years, allowing time for the development of appropriate institutional frameworks and policies. But the challenges of growth in non-communicable disease and the need for long-term care and economic security for a growing older population will also have to be confronted in regions whose public institutions and welfare regimes are barely able to cope with the current demographic profile of predominantly children and young adults. Few have the resources to face the urgent necessity of developing appropriate institutions and regimes for the 1 billion older adults who will be surviving over the next 25 years.

At the individual level, we need to consider the impact on individuals of the realization of the potential to live into the ninth or even tenth decade. As I have pointed out (Harper, 2004), all Western-style ageing societies demonstrate a clear ageing of life transitions, displaying an increase in age at first marriage and at remarriage, at leaving the parental home, and at first childbirth. Indeed, while public and legal institutions may be lowering the age threshold of full legal adulthood, individuals themselves are choosing to delay many of those transitions which demonstrate a commitment to full adulthood: full economic independence from parents, formal adult union through marriage or committed long-term cohabitation, and parenthood. It can thus be argued, for example, that because early death through disease, war, famine and (for women) reproduction is no longer the common experience, individuals feel more comfortable about establishing marital unions later in life and bearing children later.

There will, of course, be challenges as well as opportunities as we adjust to a more mature global demography. It is necessary to extend economic activity into later life, and to rethink the mechanisms of the intergenerational contract and provision of social security; throughout the world, healthcare systems will need to adjust to a reduction in acute diseases and infant and child-related medicine, and to an increase in non-communicable disease and long-term care; the bastion of institutionalized age discrimination needs to be tackled; and the reality of the experience of disease and disability at the very end of a normal lifespan needs to be acknowledged, and appropriate social care and support frameworks established. However, there are also real opportunities for a mature society: age-integrated flexible workforces, intergenerational integration, age equality and politically stable, age-integrated societies are all potential benefits of such a global demography. Benefits, however, can only be realized with the right political will across the globe.

There are indications that in both Asia and Europe, governments are grasping the challenges. In the light of Eurostat's population projections, for example, the

European Commission has identified the main areas where Member States and the Community need to act in the coming years to meet these challenges, including placing increasing productivity and encouraging late-life work at the forefront of the agenda. In Asia also, five-year planning to meet the challenges of an ageing population is underway in several countries, such as China, Singapore, Malaysia, Hong Kong and Korea.

The implications of population ageing, are, like the other challenges of the 21st century, global climate change and global terrorism, truly a global concern.

10

The Ageing Process: An Evolution in Our Understanding

Tom Kirkwood

Over the last almost 200 years, life expectancy in the countries with the longest-lived populations has increased almost linearly and without interruption. That this trend continued in recent decades is not what the forecasters predicted. Quite reasonably, they foresaw a different pattern, in which increases in life expectancy that had continued during the first half of the 20th century would begin to level off.

Their expectations were understandable. During the late 19th and early 20th centuries, crucial advances were made in standards of living through better sanitation, clean water supplies, housing, nutrition and education. These lowered rates of mortality, particularly from infections, so fewer people died young. The introduction of vaccines, followed in the middle of the 20th century by the development of antibiotics, enabled the effective control of infectious disease – previously the dominant cause of early death.

The continuing increases in life expectancy during this period were thus driven by declining mortality among those in the young and middle years of life. It was logical to assume, as did the forecasters, that once preventable causes of death had been checked, things just could not get any better.

Only the ineluctable processes of ageing would be left to dominate the patterns of morbidity and mortality, and there was little that could be done to tackle those. But, sensible though this seemed, it was wrong.

FUNDAMENTAL QUESTIONS ABOUT AGEING

In the last few decades, there has been a dramatic change in mortality patterns. The older age groups within the population have been experiencing a continuing decline in mortality, leading to a continuation in the increase in life expectancy, but now with a different driver. This raises fundamental questions about the process of ageing in our society. For example, is the nature of the ageing process itself undergoing change? Or is it that we are simply creating more benign conditions in which people can realize a greater biological potential for longevity than was possible before?

These questions create an imperative to understand the biological mechanisms of ageing. In what follows I will attempt to provide a brief overview of our current understanding.

When does ageing start?

We have very good evidence that, contrary to what some people think, ageing is not a process that begins after the age of 40, or 50 or 60. When we examine mortality rates at different ages through life, we find a significant pattern. As might be expected, mortality is significant in the very early part of life, due to the hazards of childbirth and the effects of neonatal disorders. It declines to a minimum around puberty, but then increases progressively throughout all the subsequent decades of life. The rise in mortality is not something that begins in middle or later age. It starts far earlier, during the teenage years. Ageing is a life-course process, and the importance of early-life events is reinforced by our advancing understanding of its scientific basis.

Is ageing necessary?

Most people think that ageing is, in some sense, genetically programmed into us. Underlying this idea is an implicit explanation that ageing is necessary, a process that allows each generation to make way for the next, so that the world does not become overrun with animals. At first sight, this idea seems logical. Nevertheless, it is fundamentally wrong.

That ageing may not be necessary was proved in the 1950s by a team at the Department of Zoology, Oxford University, led by David Lack. Their extensive field studies of animals in the wild demonstrated that these animals simply do not show any intrinsic sign of ageing. The reason for this is that they die young, from extrinsic hazards of the environment, long before they have a chance to become old. These findings support the work of Peter Medawar and George Williams to develop an evolutionary theory of the relationship between ageing and natural selection. This underlined the idea that ageing comes about because what happens in the late part of life operates, as far as natural selection is concerned, in a selection shadow.

Is ageing genetic?

Research has provided two crucial insights about the relationship between ageing and our genetic make-up. The first is that, biologically, there is really no potential for animals to evolve genes that actively drive the ageing process. Animals die young because the environment is dangerous. There is thus no need for an ageing process to curtail lifespan, so there are no grounds to think in terms of genetic factors that drive ageing. Furthermore, if ageing is rarely seen in nature, there cannot have been opportunity for natural selection to have given rise to a genetic programme with the specific purpose of limiting the length of life.

We can, however, go further with this line of reasoning and arrive at an understanding of how natural selection did produce the ageing process. Animal lifespans are curtailed by the high pressure of mortality imposed by the environment, and this tells us something important about the investment they are likely to be genetically predisposed to make in the metabolically expensive business of maintaining and repairing the cells and tissues of their bodies. If life really is nasty, brutish and short, then instead of struggling too hard to preserve a single body that will probably meet with a sticky end in the relatively near future, the greater priority is to reproduce, in other words to generate genetic copies to carry the genes forward into the next generation.

This is the core of the 'disposable soma' theory of ageing, which asserts that, under the pressure of natural selection, organisms are genetically tuned to put only enough investment into physical maintenance to keep the body going through the period of life expectancy that they see in the wild; more than that is a waste and will be countered by natural selection.

THE BASIC BIOLOGY OF AGEING

There are, of course, many details that need to be considered, but the following are the key insights to understanding the basic biology of ageing:

- there is no genetic programme for ageing, which is caused primarily by damage;
- there are genetic factors that regulate longevity; these are genes that set the mechanisms for maintenance and repair activities like DNA repair, antioxidant defences and protein turnover;
- ageing involves multiple mechanisms, so the biology is intrinsically complicated; and
- ageing is inherently stochastic: it is influenced by chance.

On this basis, we can build a model of what we understand the ageing process to be. It is a model that is both very simple and yet profound, describing a process in which ageing is caused by random molecular damage, which, over a lifetime, leads to an accumulation of cellular defects. This, in turn, results eventually in age-related frailty, disability and disease.

Once we have this model, we can add things that we know can exacerbate the ageing process. For example, we know that stresses of a variety of kinds – for example from adverse environments – can accelerate the accumulation of molecular damage. This is why the study of ageing is intrinsically multidisciplinary, and why it would be immensely useful to build links between those working on ageing in the biomedical sciences and those working on ageing in the social sciences, for example. Nutrition is another factor that can have adverse affects on ageing – there are many ways in which poor nutrition can accelerate the build-up of molecular damage.

On the other hand, we know of a lot of factors that help our bodily systems to correct damage, such as a healthy lifestyle and healthy nutrition. In the same way, at a higher level of function, we know that there are both inflammatory processes that result from cellular damage that can accelerate the accumulation of the diseases and frailties of old age, and anti-inflammatory factors that can be harnessed to work against these kinds of processes.

This all leads to an important conclusion: that the biological ageing process as we understand it now is intrinsically malleable. The two obvious ways to try and engender change are to decrease exposure to damage or enhance the systems that correct and repair faults. We can decrease the exposure to damage, for example, by improving nutrition, lifestyle, environment and so on. Although we need more evidence to substantiate this assertion, it seems likely that this is a large part of what has contributed to the ongoing increases in life expectancy over recent decades.

Looking forward, it may be possible to enhance the natural mechanisms for protection and repair. Some modest advances can probably be achieved in the fairly near future through better nutrition. More radical improvements may be achievable in the future through the development of novel drugs, stem cell therapies and so on.

THE BEGINNINGS OF A NEW VIEW OF AGEING

These discoveries radically alter the traditional biomedical view of ageing. We can describe this traditional view in terms of four components:

1   ageing is biologically determined, part of an inbuilt limit to lifespan;
2   ageing is a process of progressive loss and irreversible change in functional capacity;
3   ageing is a distinct degenerative phase of the life cycle; and

4   diseases of ageing are distinct from the intrinsic processes underlying healthy ageing.

The new view of age replaces these components as follows:

1   bodies are programmed for survival not death, although not – and this is the trouble – for indefinite survival;
2   the ageing process is intrinsically malleable;
3   youth and age are a continuum, which means that interventions can occur early on; and
4   intrinsic ageing and many age-related diseases share common underlying mechanisms; in fact, the best approach to tackling some of the age-related diseases is likely to be through gaining a better understanding of how intrinsic ageing works.

THE NEED FOR CAUTION

There is every reason to be very positive about the improvements in health span that may be expected from research on ageing. However, I do want to sound a very important note of caution here: despite these tremendous advances, there is a great deal more to learn. We have scarcely begun to scratch the surface of what is actually driving the ageing process.

For example, although we can modify longevity in some animal models, the relevance of this to human health and health-span extension is uncertain. Short-lived animals have evolved a considerable capacity for flexibility in their life histories, in order to cope with unpredictable environments. Indeed, most of the capacity for lifespan extension in these animal models exploits pre-existing flexibility, so it is unsurprising that we can easily extend their lifespans. However, we have very little evidence for the efficacy of any comparable intervention to extend human longevity or health span.

Some people have raised the possibility of harnessing gene therapy to bring about radical changes in human longevity. Perhaps this will be possible in the future, but at the present time, it is worth remembering that it is not yet possible to perform successful gene therapy even for the most straightforward of targets, such as single-gene disorders like cystic fibrosis. Moreover, many of the claims being made in the stem cell area fail to take account of enormous obstacles which must be overcome before therapies for age-related diseases can be delivered.

In summary, the potential future benefits of life-extension technologies and their possible timescales remain largely a matter of conjecture at this point. But this does not stop us from discussing them. Discussion of life extension is now commonplace, but rather worryingly it seems to involve a much more lax approach to the requirement for caution than is the case in other, better-established spheres of biomedical endeavour. It may be that this in turns reflects society's ambivalent views about the wider aspects of ageing, where fantasy has long exercised an unusual fascination.

CHALLENGES AND OPPORTUNITIES, EDUCATION AND ENGAGEMENT

My final observations concern approaches to public education and engagement with regard to the challenges of population ageing. It is important that researchers explain the current science to the wider public, given that this is an area of such pervasive relevance for all our lives. This is a crucial part of persuading people that ageing is a tractable problem deserving of investment. It will enable us to expand our research capacity, to recruit more clinicians to address the medicine of old age and to involve industry. It is also imperative that we engage government – and that means discussing the challenges that exist, honestly and sensibly. In this regard, I would like to close by drawing attention to the wide-ranging report on 'Ageing: Scientific aspects' from the House of Lords Science and Technology Select Committee (House of Lords, 2005), which sets out some of the challenges for all those who need to be involved.

11

Postponing Ageing:
Re-identifying the Experts

Aubrey de Grey

SENS: The Early Years

In July 2000, having worked in biogerontology for about five years, I realized that the field might collectively be making a major oversight with regard to the available options for postponing ageing substantially in mammals and eventually in humans. In a nutshell, I wondered whether the dictum that 'prevention is better than cure' was being taken too far (see Figure 11.1).

Ageing is a side-effect of metabolism: the functional decline of old age results from the eventually intolerable accumulation of adventitious molecular and cellular side-effects of intrinsic and essential metabolic processes. It is clear that the geriatrics approach to intervention is a losing battle, because these side-effects continue to accumulate and the task of maintaining health becomes

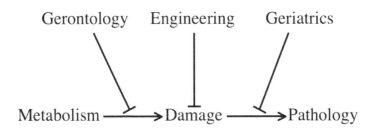

Figure 11.1 *How the engineering (SENS) approach to postponing ageing relates to the two traditional approaches*

progressively harder. However, the fact that ageing usually has no great functional consequences until the second half of life tells us that there is a threshold level of these accumulating side-effects, below which they are essentially harmless. Thus, in principle, one might be able to determine the set of eventually pathogenic differences in composition between young and middle-aged adults – differences that I collectively refer to as 'damage' – and find ways to reverse those differences, so restoring the middle-aged individual to a more youthful state. I saw a major potential attraction of this 'rejuvenation' approach: the initially inert character of damage means that eliminating it is likely to be much easier than preventing it from being laid down in the first place (which, essentially, is the alternative historically favoured by biogerontologists), because the latter involves manipulating bioactive targets that are participating in a process that we still understand extremely poorly, namely our metabolism. I came to the conclusion that the types of damage could indeed be quite comprehensively enumerated and, moreover, that feasible ways either to reverse ('repair') or to obviate each of them were either already in existence in mice or else built on sufficiently detailed *in vitro* work that their development in mice could reasonably be expected within a decade (see Table 11.1).

I stress the word 'might' in the first sentence of this chapter. As a theoretician, with no first-hand experience of experimental biology, I am always acutely aware that I may underestimate the difficulty of future work. Consequently, I at once convened a one-day workshop to analyse my proposal in detail; this occurred in October 2000 and led to an article co-authored by all the participants (de Grey et al, 2002). I have continued this approach since then, running three further workshops and three full-scale conferences. These and my related activities have collectively become known as the SENS project, standing for 'Strategies for Engineered Negligible Senescence'. This highly provocative moniker is appropriate because I claim that, once we achieve sufficient progress in these technologies to give middle-aged people an extra 30 years of healthy life (roughly doubling their remaining lifespan), our rate of progress in further refining those therapies is very

Table 11.1 *Foreseeable approaches to repair or obviation of the seven major types of damage that accumulate during mammalian ageing*

Type of damage	Proposed repair (or obviation)
Cell loss, cell atrophy	Stem cells, growth factors, exercise
Death-resistant cells	Ablation of unwanted cells
Oncogenic nuclear mutations/ epimutations	'WILT' (whole-body interdiction of lengthening of telomeres)
Mitochondrial mutations	Allotopic expression of 13 proteins
Intracellular aggregates	Microbial hydrolases
Extracellular aggregates	Immune-mediated phagocytosis
Extracellular cross-links	AGE-breaking molecules (inhibitors of advanced glycation endproducts)

likely to exceed the rate at which the remaining imperfections in the therapies are accumulating in the bodies of those receiving the latest treatments. In other words, people will become biologically younger as they become chronologically older, eventually becoming permanently as youthful as today's 20 to 30 year-olds, subject only to the periodic application of these therapies (de Grey, 2004).

Attention to my proposals within the biogerontology community was muted for as long as it could be. Such extreme predictions are not conducive to a quiet life for the field and were rapidly feared to be in danger of bringing biogerontology into disrepute; this indeed hindered (though never prevented) publication of some of the articles arising from my workshops, despite the eminence of their co-authors. Beginning in 2003, however, the mainstream media gradually became aware of SENS, and senior biogerontologists found themselves having to respond to questions about it from journalists. At that point, I am afraid to say that, in what I can only interpret as a reaction to a perceived threat to their authority, some of my colleagues began to resort – off the record, in the main – to the undeniably *ad hominem* argument that I could be safely ignored because I have no experimental training. I found this particularly saddening, because those same colleagues had admirably desisted from such a stance during my first few years in the field, when I was presenting highly novel but politically harmless ideas; those ideas had always been judged purely on their merits.

SENS's Advance to 'Gandhi Stage 3'

A famous aphorism from Mahatma Gandhi concerning the progression of the mainstream's reaction to radical new ideas goes roughly as follows: first they ignore you, then they laugh at you, then they oppose you, then they say they were with you all along. It will perhaps be seen that the reaction to SENS outlined above rather accurately fits the first two of these stages. Aware that only after advancing to stage three would SENS be subjected to the detailed critical analysis that it must experience and withstand in order to achieve the credibility needed for adequate funding, I became increasingly vocal during 2004 and 2005 in deprecating my critics' failure to discuss SENS seriously. With the help of a sharp acceleration during the same period in media attention to SENS, this effort succeeded quite spectacularly in late 2005. In July 2005, I published a commentary noting biogerontologists' public duty to forsake dogma in the interests of finding a solution to ageing as soon as possible and thus saving staggering numbers of lives, and I noted several ways in which the field was departing from this imperative (de Grey, 2005a). One colleague whom I criticized in this piece, Richard Miller of the University of Michigan, responded with a rejection of SENS's feasibility – this was, however, lacking in defence of the aspects of his and other senior biogerontologists' preferred research directions and rhetoric that I had so strongly critiqued (Warner et al, 2005). What I had not predicted was that, rather than replying on his own,

Miller amassed an entourage of fully 27 co-signatories (one of whom, Huber Warner, agreed to be the lead author) spanning essentially all specialities within biogerontology.

It may not be instantly apparent that this was, from my point of view, a vastly more favourable outcome than if Miller had simply responded as a sole author. What made it so was that this first attempt by SENS's detractors to justify their stance with concrete arguments, based on published data, was – as I of course expected, but I might have been wrong – flawed on numerous levels, as I was able to explain in my response published in the same issue of *EMBO Reports* (de Grey, 2005b). Since the general public remains obstinately ambivalent concerning the desirability of postponing ageing – something that I claimed in the original *EMBO Reports* commentary is largely the fault of biogerontologists' fixation on political correctness – there is essentially no prospect of public funding for SENS, so philanthropy is my main focus. From the point of view of a potential benefactor, the fact that one professor bases his low opinion of SENS on flawed logic does not mean that SENS is necessarily sensible. When this flawed critique emanates from a host of experts spanning the whole field, however, it would appear that biogerontologists in general are simply not equipped to opine authoritatively on the overall SENS programme: their views should be taken into account, to be sure, but only in conjunction with the views of experts in the many other biological disciplines that SENS has brought together. Accordingly, SENS's prospects at the due diligence stage are vastly enhanced by the existence of incontrovertible documentary evidence showing that those upon whom one might imagine one could rely for expert evaluation of such ideas are in fact in possession only of part of the necessary expertise.

THEORETICIANS IN BIOLOGY: TIME-WASTERS OR ASSETS?

There is a more general point to be made here. In physics, a symbiosis exists between those who toil at the bench, telescope and so on and those who occupy their time pacing up and down and talking to themselves. This is productive, because cutting-edge experimental work is extremely time-consuming: one simply cannot generate important data without becoming rather narrowly specialized and avoiding the temptation to read widely beyond one's area, but the synthesis of results and ideas from ostensibly distant subfields within physics regularly leads to new insights concerning what experiments are and are not worth doing next. The partitioning of physicists into theoreticians and experimentalists is thus a rational division of labour that has amply stood the test of time. The existence of whole departments of theoretical physics is testament to this.

The contrast with biology could hardly be starker. There are two quite large groups of biologists who tend to be described as theoreticians, but they are not counterparts of theoretical physicists at all. One is the evolutionary biologists, who are primarily mathematicians working with a small and sharply circumscribed subset of biological data. The other is the computational biologists, who are mainly concerned with simulating biological processes *in silico* – an activity that is readily understood by those (including myself, fortuitously) who have worked in experimental computer science to be simply a different type of experiment. Theoreticians of the type I have described above in physics are virtually unknown in biology. This is all the more paradoxical given the rather impressive hit rate achieved by the few counter-examples: the proportion of Nobel laureate biologists who lacked experimental training exceeds by a large factor the incidence of such individuals in biology as a whole.

The above statistic is but one way in which it can be seen that biology would probably benefit a great deal from at least partially following physics's lead and encouraging more students to pursue a degree of cross-disciplinary expertise that is essentially incompatible with success in experimental science. Another argument is simply that science is science is science: any field too large to be mastered in one's spare time is too large to be populated entirely by experimentalists if it is to progress at the maximum possible rate.

Why, then, have experimental biologists fallen so disastrously prey to the prejudice that only those with experimental training are competent to identify which experiments are worth doing? I cannot, in all honesty, claim that it is entirely their fault. It is a sad fact that an immense amount of absolutely worthless theoretical biology has been published over the years by newcomers to the field who have not been fortunate (or, less charitably, careful) enough to study the published literature thoroughly before rushing into print with their beautiful new explanation of a small subset of relevant experimental data. While this is actually quite useful to those who lack experimental training but distinguish themselves by their care to avoid the above error – I, for example, rose unusually rapidly to be respected as an intellectual equal by my senior colleagues, since the fact that I could have valuable ideas despite my educational 'handicap' was a source of esteem – it clearly does theoretical biology in general no good at all.

Experimental biologists should, I thus contend, make more effort to sort the wheat from the chaff of theoretical biology. I succeeded not only because I was careful, but also because, mainly through sheer luck, I was in a position to attend a great many conferences and become known personally as having a quick mind and useful things to say about others' work. This degree of face-to-face interaction with those who currently define my field of interest is a luxury not enjoyed by most newcomers.

CONCLUSION: DEFEATING AGEING NEEDS A VILLAGE

In closing, I return to my main topic: the effort to combat the affliction that kills at least twice as many people worldwide (and perhaps 10 times as many in the West) as all other causes combined, namely ageing. I hope I have demonstrated that the resources relevant to this struggle have not been optimally marshalled, on account of the understandable bias in favour of their own work that characterizes the views of those whom the public and policymakers have innocently, but mistakenly, regarded as uniquely authoritative concerning the promise of various approaches to the problem. The brouhaha over SENS summarized in this chapter was not something that I remotely enjoyed, involving me as it did in the strident criticism of, and receipt of the same from, good friends; accordingly, my relief that it was a rather brief episode arises not only from the consequent acceleration of funding and progress that SENS has since enjoyed, but also because I have been able to revert to entirely cordial relations with these colleagues. However, I could not let such emotions divert me from a course that will certainly save countless lives in the future. I say 'certainly', because the decisive rejection of SENS, if that is the eventual outcome of the experimental work that gerontologists and (predominantly, in fact) others are now starting to undertake, will clear the way for the introduction of new and better ideas, just as early success in implementing parts of SENS will hasten its full implementation. In other words, whether SENS is sense or nonsense, we are now set to defeat ageing sooner than if SENS had continued in the state of unevaluated limbo that it occupied initially.

12

In Pursuit of the Longevity Dividend

S. Jay Olshansky, Daniel Perry, Richard A. Miller and Robert A. Butler

Imagine an intervention, such as a pill, that could significantly reduce your risk of cancer. Who wouldn't jump at the chance to take it? Imagine an intervention that could significantly reduce your risk of stroke, or dementia, or arthritis. It would be a market blockbuster over night. Now imagine an intervention that does both of these things, and at the same time also reduces your risk of everything else undesirable about growing older, including heart disease, diabetes, Alzheimer's and Parkinson's disease, hip fractures, osteoporosis, sensory impairments, and sexual dysfunction; in fact, everything that goes wrong with us as we age. Such a pill may sound like a fantasy, but ageing interventions in animal models already do just this, and many biogerontologists believe such an intervention is a realistically achievable goal for people, too. We already place a high value on both quality and length of life, which is why we try to immunize all children against infectious diseases. In the same spirit, we suggest that a concerted effort to slow ageing should be sought after for the same reasons – because it will save and extend lives, improve health, and create wealth.

The experience of ageing is about to change. Human beings are approaching old age in unprecedented numbers, and this generation and all that follow have the potential to live longer, healthier lives than any in history. But the changing demographics of our species also carry the prospect of overwhelming increases in the prevalence of age-related disease, frailty, disability, and all the associated costs and social burdens. The choices we make now in the public policy arena will have a profound influence on the health and wealth of current and future generations as well as the global economy.

GERONTOLOGY COMES OF AGE

Gerontology has grown far beyond its historical and traditional image, of disease management and palliative care for the old, to the scientific study of the processes of ageing in humans and other species – the latter known as biogerontology. In the past few decades biogerontologists have gained significant new insights into the causes of ageing; made discoveries that have revolutionized our understanding of the biology of life and death; destroyed long-held myths about ageing and the aged; and both gained public trust and ignited once again the long-held interest in extending the length and improving the quality of our lives. For example, the old idea that each illness expressed at older ages was independently influenced by genes and/or behavioural risk factors has been dispelled by evidence that genetic and dietary interventions can retard nearly all aspects of late life-diseases in parallel. Genes and hormones that extend life in simpler organisms have been found, to everyone's surprise, to do so in mice and rats as well, in other words in animals in which ageing leads to most of the same trials and tribulations that people experience as they age. These findings suggest that our own bodies, like those of laboratory animals, may well have 'switches' whose position modulates how quickly we age. These switches that influence ageing are not set in stone, but potentially adjustable.

Biogerontologists have progressed far beyond merely describing cellular ageing, cell death, free radicals and telomere shortening, to the point at which they might manipulate and alter molecular machinery and cell functions (Warner, 2005). These recent scientific breakthroughs made by committed biogerontology researchers have nothing in common with the claims of prominent present-day entrepreneurs out to make money by selling alleged anti-ageing interventions they say can already slow, stop, or reverse human ageing. More importantly, however, the long-standing belief that ageing is an immutable process that was programmed into us by evolution is now known to be wrong.

Today's biogerontologists routinely extend the period of healthy life in many kinds of laboratory animals through genetic or nutritional means. The scientists involved in this exciting new line of research are embedded in the most prestigious universities and medical schools, and publish in the best peer-reviewed journals. Our knowledge of how, why and when the processes of ageing take place has progressed so much in recent decades that many scientists now believe that people alive today could benefit from advances in this line of research if it is sufficiently promoted (Miller, 2002; Olshansky et al, 2002; Public Agenda, 2005). Indeed, the science of ageing has the potential to do what no drug, surgical procedure or behaviour modification can do – extend our years of youthful vigour and simultaneously postpone all of the costly, disabling and lethal conditions expressed at later ages.

The economic benefits to individuals and to nations that would accrue from the extension of healthy life would be enormous. By extending the duration of time in the lifespan when higher levels of physical and mental capacity are expressed, people would remain in the labour force longer, personal income and savings would be enhanced, age-entitlement programmes would face considerably less pressure from shifting demographics, and there is reason to believe that national economies would flourish. The science of ageing has the potential to produce what we refer to as a 'Longevity Dividend' in the form of social, economic and health bonuses both for individuals and the populations they comprise – a dividend that would begin with generations currently alive and continue from one generation to the next.

We contend that conditions are ripe today for the aggressive pursuit of the Longevity Dividend by seeking the technical means to intervene in the biological processes of ageing in our species, and to ensure that the resulting interventions become widely available to the entire population.

Why Act Now?

Consider what is likely to happen if we choose not to. Take, for instance, the impact of just one disorder associated with growing older – Alzheimer's disease (AD). For no other reason than the inevitable shifting demographics, the number of Americans stricken with AD is expected to rise from 4 million today to as high as 16 million by the middle of this century (Liesi et al, 2003). This means there will be more people with AD in the US by 2050 than the entire current population of Australia. Globally, the prevalence of AD is expected to rise to 45 million by the year 2050, with three of every four people afflicted by the disease living in a developing nation (Alzheimer's Disease Annual Report, 2004–2005). The US economic toll is currently pegged at $80–$100 billion, but by 2050 more than $1 trillion will be spent annually on AD and related dementias. The prevalence of this single age-related disease will be catastrophic, and this is just one example.

Cardiovascular disease, diabetes, cancer and a host of other costly age-related health problems account for countless billions of dollars siphoned away for 'sick care'. In China and India, the elderly population, by mid-century, will outnumber the total current US population. And in many developing nations there is little or no formal training in geriatric healthcare. The demographic wave is a global phenomenon that appears to be leading healthcare financing into an abyss.

Nations may be tempted to continue attacking the diseases and disabilities of old age separately, as if they were unrelated to one another. This is the way most medicine is practised and medical research conducted today – one disease at a time. The National Institutes of Health (NIH) in the US, for example, are organized under the premise that specific diseases and disorders should be attacked individually, in isolation, and this is reflected in separate institutes created for each major disease. In fact, more than half of the National Institute on Ageing (NIA) budget in the US is devoted to AD. Government and non-government organizations in other countries are similarly structured. But the underlying biological changes that predispose everyone to fatal and disabling diseases and disorders are caused by the processes of ageing (Butler et al, 2004). It therefore stands to reason that an intervention that delays ageing should become one of our highest research priorities.

HEALTH AND LONGEVITY CREATE WEALTH

The benefits of delayed ageing surpass the extension of healthy life. According to studies undertaken at the International Longevity Center and a number of universities around the world, the extension of healthy life creates wealth for individuals and the nations in which they live (Bloom and Canning, 2000). Healthy older individuals accumulate more savings and investments than those beset by illness; healthy older people tend to remain productively engaged in society, both in compensated work and in community service and civic engagement; they are well known to spark economic booms in the so-called mature markets, including financial services, travel, hospitality and intergenerational transfers to younger generations; improved health status leads to less absenteeism from school as a child and work as an adult; and healthier people tend to be better educated and receive higher pay and more income than less healthy people throughout the life course.

A successful intervention that delays ageing would do more than yield a one-off financial and health benefit, after which, one might argue, the same high healthcare expenses would ensue. Life extension already achieved among animals suggests that delayed ageing may produce a genuine compression of mortality and morbidity (Vergara et al, 2004) – a scenario that, if it applied to people, would yield substantial permanent reductions in healthcare costs. Animals that have been calorie restricted not only experience a reduction in their risk of death, they also experience declines in the risk of a wide variety of age-sensitive, non-lethal conditions such as cataracts, kidney diseases, arthritis, cognitive decline, collagen cross-linking and immune senescence (Miller and Austad, 2006). If this could be achieved in people, the benefits to health and vitality would begin immediately with the intervention and continue throughout the remainder of the lifespan. Thus the costly period of frailty and disability would be experienced during a shorter duration of time before death. Just like vaccinations, which are desirable because they protect the health of every generation of children, delayed ageing would provide a second highly desirable gift of extended health to these same generations as they reach middle and older ages.

This compression of mortality and morbidity would *create* financial gains not only because ageing populations will have more years to contribute, but also because there will be more years during which age-entitlement and healthcare programmes are not utilized.

A RIPENING SCIENCE

Scientists have believed for centuries that ageing exhibits common characteristics across species, and recent work in genetics and in the comparative biology of ageing has not only confirmed these early impressions but also provided important clues about how to develop effective interventions that delay ageing. It is now clear that some of the hormones and cellular pathways that influence the rate of ageing in lower organisms also contribute to many of the manifestations of ageing that we see in humans, such as cancers, cataracts, heart disease, arthritis and cognitive declines. These manifestations occur in much the same way in other animals and for the same biological reasons (Sinclair and Guarente, 2006). Several experiments have demonstrated that by manipulating certain genes, altering reproduction, reducing caloric intake and changing the signalling pathways of specific physiological mechanisms, the duration of life of both invertebrates and mammals can be extended (Weindruch and Sohal, 1997; Tatar et al, 2003). Some of the genes involved, such as Pit1, Prop1 and GHR/BP, modulate the levels of hormones that affect growth and maturation; others, such as p66-Shc, help individual cells avoid injury and death. No one is suggesting that alteration of these genes in humans would be practical, useful or ethical, but it does seem likely that further investigation of how these genes prevent diseases and extend healthy lifespan may yield important clues about how to accomplish similar things pharmacologically.

Genes that slow growth in early life – such as those that produce differences between large, middle-size and miniature dogs – typically postpone all the signs and symptoms of ageing in parallel. A similar set of hormonal signals, related in sequence and action to human insulin, insulin-like growth factor (IGF-I) or both, are involved in ageing, lifespan and protection against injury in worms, flies and mice, and extend lifespan in all of these animals. These hormones operate at the cellular level, helping individual cells throughout the body buffer the toxic effects of free radicals, radiation damage, environmental toxins and protein aggregates that contribute to various late-life malfunctions.

An extension of disease-free lifespan of approximately 40 per cent has already been achieved repeatedly in experiments with mice and rats (Yu et al, 1985; Weindruch and Walford, 1988; Brown-Borg et al, 1996; Flurkey et al, 2001). These examples provide powerful new systems to study how ageing processes influence disease expression and will yield important clues about where to look for drugs and nutritional interventions that can eventually be used to slow down ageing in people in a safe and effective way. Since many of ageing's biological pathways are conserved also in simple invertebrate species such as fruit flies, it should be possible to experimentally evaluate candidate intervention strategies rapidly.

It is also known that some people, including a proportion of centenarians, live most of their lives free from frailty and disability. Genetics play a critical role in their healthy survival. Identifying variation in these subgroups of human beings holds great potential for improving public health among the majority of the population. For example, microsomal transfer protein (MTP) on chromosome 4 has been identified as a longevity modifier in a sample of centenarians (Geesaman et al, 2003); there is strong evidence linking a common variant of *KLOTHO*, the KL-VS allele, to human longevity (Arking et al, 2005); and it has been demonstrated that lipoprotein particle sizes promote a healthy ageing phenotype through codon 405 valine variation in the cholesteryl ester transfer protein (CETP) gene (Barzilai et al, 2003).

Given the speed with which advances in the scientific study of ageing have already occurred, and our ability to obtain research results quickly from the study of short-lived species, scientists have reason to be confident that a Longevity Dividend is a plausible outcome of ageing research.

THE TARGET

What we have in mind is not the unrealistic pursuit of dramatic increases in life expectancy in the order of hundreds or thousands of years, let alone the kind of biological immortality best left to science fiction novels (Warner et al, 2005), but a goal we believe is realistically achievable: a modest deceleration in the rate of ageing sufficient to delay all ageing-related diseases and disorders and their attendant costs by about seven years (Olshansky, 2003). This target was chosen because the risk of death and most other negative attributes of ageing that are associated with the passage of time tend to rise exponentially throughout the adult lifespan, with a doubling time of approximately seven years (Butler and Brody, 1995). Such a delay would yield health and longevity benefits greater than would be achieved by the elimination of any single fatal disease such as cancer or heart disease (Olshansky, 1987). And we believe it can be achieved for generations now alive.

If we succeed in slowing ageing by seven years, the age-specific risk of all causes of death, frailty and disability will be reduced by approximately half at every age. Equally important, once achieved, this seven-year delay would yield equal health and longevity benefits for all subsequent generations as they pass through the lifespan in much the same way that children born in most nations today benefited from the discovery and development of immunizations in the past two centuries.

What this means is that people who reach 50 years of age in the future would have the health profile and disease risk of today's 43 year old, those aged 60 would resemble current 53 year olds, and so on. Although the health benefits of delaying individual degenerative diseases have been well established (Olshansky et al, 2005), an intervention that slows ageing would not merely enable people to live longer, but would be expected to extend the physical, physiological and psychological vigour of youth. Our expectation is based on a careful and conservative evaluation of the results of animal experiments in which both lifespan and the period of physiological and cognitive vigour of youth have been jointly extended by nutritional or genetic interventions.

A growing chorus of leading scientists agree that this objective is scientifically and technologically feasible. How quickly we see success is dependent in part on the priority and level of support devoted to the effort. Certainly such a great goal – to win back an average of seven years of healthy life – deserves and requires significant resources in time, talent and treasury. Compared to the mammoth investment already committed by both the public and private sectors to caring for the sick as they age, and the pursuit of ever more expensive treatments and surgical procedures for existing fatal and disabling diseases, the pursuit of the Longevity Dividend would be modest, and the pay-off, in terms of both health and economics, would be far greater than curing any single major fatal disease. In fact, an investment in the Longevity Dividend would probably pay for itself, because a healthier, longer-lived population will *add* significant wealth to the economy, not take it away.

THE RECOMMENDATION

The National Institutes of Health were funded at $28 billion in 2006, but less than 0.1 per cent of that amount goes to understanding the biology of ageing and how it predisposes us to a suite of costly diseases and disorders expressed at later ages. We are calling on Congress to invest 3 billion dollars annually to this effort, or about 1 per cent of the current Medicare budget of $309 billion; to provide the organizational and intellectual infrastructure and other related resources to make this work; and make a commitment to this effort like that made by President Kennedy to land astronauts on the moon and return them safely.

Specifically, we recommend that one third of this budget ($1 billion) be devoted to the basic biology of ageing, with a focus on genomics and regenerative medicine as they relate to longevity science. Another third should be devoted to age-related diseases as part of a coordinated trans-NIH effort. One sixth ($500 million) should be devoted to clinical trials, with proportionate representation of older persons (aged 65+), that include head-to-head studies of drugs or interventions, including lifestyle comparisons, cost-effectiveness studies and the development of a national system for post-marketing surveillance. The remaining $500 million should go to a national preventive medicine research initiative. This initiative would include studies of safety and health in the home and workplace and address issues of physical inactivity and obesity as well as genetic and other early-life pathological influences. This last category would include studies of the social and economic means to effect positive changes in health behaviours in the face of current health crises – obesity and diabetes – that can lower life expectancy (Olshansky et al, 2005). Elements of the budget could be phased in over time, and it would be appropriate to use funds within each category for research training and the development of appropriate infrastructure. We also strongly encourage the development of an international consortium devoted to this task, as all nations would benefit from securing the Longevity Dividend.

With the effort we propose, we believe it would be possible to intervene in ageing among the baby boom cohorts, and certainly all generations after them would enjoy the health and economic benefits of delayed ageing. Such a monetary commitment would be small compared to what we already spend each year on Medicare alone, but it would pay dividends an order of magnitude greater than the investment, and it would do so for all future generations.

Our vision of ageing in the 21st century is one of optimism. Although it is easy to be alarmed about what might come to pass if we do not actively pursue efforts to slow the biological processes of ageing, we choose instead to imagine how public health and wealth would benefit if we did. In our view, the scientific evidence strongly supports the idea that the time has arrived to invest in the future of humanity by encouraging the commensurate political will, public support and resources required to slow ageing, and to seize this opportunity now so that most people currently alive might expect to benefit from such an investment. A successful effort to extend healthy life by slowing ageing may very well be one of the most important gifts that our generation can give to all current and future generations.

13

From Ageing Research to Preventive Medicine: Pathways and Obstacles

Richard A. Miller

Those charged with tracing out a 'map to the future' confront an early branch point: either to select the well-travelled and alluring road to science fiction, in which magical pills stuffed with telomeres and stem cells make us immortal overnight, or to appeal instead to readers who, buoyed by curiosity and optimism but balanced by at least a remnant of critical sense, would prefer to follow pathways that connect the known to the plausible. This chapter will take the more pedestrian fork, wearing with a quiet pride a lapel pin proclaiming membership of the reality-based community (Warner et al, 2005).

In the mid-1930s, cancer biologists made an accidental discovery that should have revolutionized cancer biology, but hasn't yet. They tested the idea that diminished calorie supplies would diminish the rate of spontaneous cancers. They found that rodents allowed to eat only about 60 per cent of the amount of food they would choose to eat if they were given free rein, had a dramatically reduced rate of cancer, and in fact lived longer than control animals. Follow-up studies, many in the 1970s and 1980s, showed that this phenomenon was routinely reproducible in mice, rats and indeed virtually every other species examined. These follow-up studies also led to the remarkable discovery that this caloric restriction (CR) intervention not only postponed virtually every kind of spontaneous cancer, but also led to a parallel delay or deceleration in nearly all other aspects of ageing. The CR diet slowed the rate of development of kidney disease, arthritis, immune senescence, cognitive decline, autoimmune illnesses, cataracts and virtually all other forms of late-life pathology. It slowed age-related changes in cells that divided regularly, cells that divided only rarely, cells that never divided and materials outside of cells altogether. The animals on the experimental diet remained not only alive, but healthy and vigorous to ages far beyond those normally seen in control populations (Weindruch and Sohal, 1997).

The size of the effect seen is dramatic: in most studies, an increase of 30–40 per cent – occasionally more – in average and maximal longevity. To put this into context, demographic calculations (Olshansky et al, 1990) show that the complete conquest of either cancer or heart disease would lead to an extension of average human lifespan by about 3 per cent in a modern industrialized society. Most readers of this chapter will know that preventive approaches that lead to the complete abolition of cancer or heart attacks are not yet available even for laboratory animals, and are not likely to be for some time. How is it, then, that a method that is 10 times *more* effective, in terms of extension of healthy lifespan, has so far failed to make much of an impact on the global research enterprise? (And fail it has – of every $10,000 spent by the US National Institutes of Health, for example, only about $6 is invested into research into the fundamental biology of the ageing process and its controls.) Calorie restriction is not likely to be useful as a direct approach to prevention in humans, because very few people can endure the extreme degree of restriction needed. Given strong motivators, including avoidance of disease and social stigma, most people can lose a few per cent of their body weight, but only a small fraction of these maintain weight loss for a year or more; the CR diet leads to a loss of about 40 per cent of body weight, so that an 80kg man, for example, would have to lose 32kg and then maintain that weight indefinitely. Thus CR diets are not a realistic treatment option, but it seems entirely plausible that a detailed analysis of the biochemical, metabolic and hormonal changes induced by CR would provide multiple new leads to the biology of ageing, towards the goal of developing pharmaceuticals that induce similar changes in young and middle-aged adults. In view of the exceptionally large effects of CR on so many of the diseases and disabilities of ageing, why is so little effort going into this area of research? Why isn't an analysis of the basic biology of ageing, its synchrony, its control and its relationship to disease the central focus of modern medical research, as indeed it should be?

The problem, I think, is largely social, a matter of public perception, or rather misperception, with the guilt equally distributed among the scientists, administrators and legislators who direct and control spending on science and those teachers and journalists from whom the public develops attitudes and preferences about research agendas. Many of the obstacles blocking pursuit of biogerontology as the quickest and cheapest path towards effective preventive medicine are also 'systems level' problems, and hence eminently appropriate topics for discussion among experts in economics, social systems and governmental affairs. Some of these problems are summarized below.

1   A major problem is that the study of anti-ageing interventions is highly stigmatized, its adherents often viewed as an undifferentiated mass of quacks and con artists. Serious scientists know that effective anti-ageing medicines are still hypothetical, that there is not yet any compelling evidence that ageing in humans (or even rodents) can be postponed by a drug or single dietary supplement. But wishful thinking by gullible customers, combined with a plentiful supply of avaricious and unscrupulous promoters, in the absence of effective consumer protection laws, leads to the sale of $10 billion annually of unproven, useless or dangerous 'anti-ageing' drugs in the US alone (Eisenberg et al, 1993). Scientists and administrators, eager to show that they have not been taken in by such hucksters, too often fail to discriminate hype from genuine progress, flashy sales pitches from well-considered research strategies, and consider the entire research area to be tainted and ill-founded. Journalists eager to attract readers, and conference organizers eager to attract journalists, too often show a distinct preference for speakers whose confidence and celebrity outweigh their credibility and critical sense; such affinity for the outré continues to obscure public awareness of the secure foundation and potential importance of modern biogerontological research.

2   A second set of problems is economic: the development and testing of authentic anti-ageing medicines would be costly and take many years. Pharmaceutical firms, and those who run them, are driven by a different pacemaker, in which the company's fortunes and those of its directors depend upon the development of patentable, high-selling drugs at short intervals – every year or two. There is little or no motivation, in such a setting, for evaluating an agent that might have to be given to healthy young adults for decades before its effects were seen – even if the effects were as dramatic as those seen in calorically restricted mice, even if preclinical trials of the agent in rodent models had generated extremely promising data. An efficacy trial of such an agent, no matter how well it worked, would be difficult to manage and costly to conduct, and, worst of all, would produce profits only for executives yet unborn (or at any rate unfledged and unannointed). Start-up companies, hoping to mollify the intense pressure of investors for quick returns, typically develop business plans that feature introduction of candidate agents for other indications, with the not-so-secret hope that once introduced into clinical practice they might then be extensively used for other 'off-label' indications, such as delay of the ageing process. Many of these firms, meanwhile, rake substantial profits from the promotion of non-prescription drugs (excuse me, 'nutritional supplements') whose effects on ageing are supported more by winks and nods than by evidence. In a system where negative evidence will hurt sales, and pursuit of authentic anti-ageing drugs would be risky, costly and very slow, it is hard to see why a responsible pharmaceutical firm would have much interest in ageing research.

3   In the public arena, a combination of forces conspire to make vigorous support for authentic anti-ageing research far too risky for most public figures. Part

of the problem is ignorance: politicians and scientific administrators who understand that research on the ageing process is likely to be the most effective, quickest and least costly path to prevention of important late-life diseases are still lamentably rare. Part of this is stigma: arguing that research to slow ageing is practical and important brands one as (apparently) unsound, apparently as someone who cannot tell the difference between science fiction and real discovery, someone apparently easy to fool. And part is competition: lobbyists for research on breast cancer, Parkinson's disease, macular degeneration, AIDS, Alzheimer's disease and dozens of other important illnesses are numerous, well funded, well connected and ruthless in their pursuit of a maximized share of the public and philanthropic purses. These lobbyists play hardball, armed as they are with authentic sufferers, friends of sufferers and sometimes recovered former sufferers, and they make expert appeals to voters and legislators, many of whom know someone with the specific diseases in question. They are generally uninterested in the suggestion that work on the biology of ageing – in other words on the underlying cause not just of their disease of interest but of most other diseases as well – may be a surer path to prevention than the disease-specific research programmes favoured by the traditional medical establishment.

4   A fourth set of problems reflects the career choices of researchers themselves. Smart young researchers, not surprisingly, prefer to work in areas where there is sufficient financial support to pay for salaries, assistants and research *matériel*. Scientists without support do not get publications; those without publications do not get grants; those without grants do not get (or at any rate retain) jobs. Thus when the US National Institute on Ageing decides, upon Congressional compulsion, to devote half of its budget to Alzheimer's research, it is not surprising to find many of the best scientists, both senior and fledgling, acquire a sudden intense interest in neuropathology and the chemistry of amyloid. Junior scientists can sometimes be lured into relatively obscure research areas by the promise that they will learn new and recondite methods, or get to work with large, expensive equipment. Ageing research, in contrast, has little to offer in the way of new toys: there is very little visceral thrill associated with experiments that depend largely on the enumeration of thousands or tens of thousands of dead flies, worms or mice.

5   The final entry on this list is the common, though indefensible and irrational, syndrome of 'gerontologiphobia', defined as a fear that biogerontologists may someday learn enough about ageing to figure out how to delay late-life diseases in real people. Gerontologiphobics envision a dystopia in which a tiny minority of healthy young and middle-aged people toil endlessly to support the lazy post-retirement lives of their parents, grandparents and so forth, all of these ancestors assumed to be arthritic, semi-demented and crotchety to boot. These worrywarts fail to understand that delay of ageing extends the duration of healthy, active vigour. This assertion is not mere guesswork, but is based on

consistent evidence; at their (late) death, calorically restricted and slow-ageing mutant mice show fewer signs of disease and disability than their long-dead control cousins did at their own, earlier, demise. Preventive strategies that simply prolong the years of incapacity, pain and dependence would be hard to justify and presumably unpopular. Fortunately, the anti-ageing manoeuvres that work in laboratories, and which might lead to workable preventive strategies for people, prolong health rather than illness.

A second motif is Malthusian: justifiably concerned about the Earth's dwindling resources and the perils of population expansion, the gerontologiphobics view ageing research as the surest way to use up the last drop of oil, the last grain of rice and that last open parking space. It is indeed hard to overstate the horrors likely to follow in the next few decades unless politicians somehow, against all present odds, face up to their responsibilities to recognize and confront resource depletion and the consequences of religious opposition to birth control. But in fairness it can be pointed out that biogerontologists, who have yet to discover anything of use to anyone, have not contributed one iota to the present, looming mess, so that obsessing over the evil effects of hypothetical anti-ageing interventions betrays a certain distractibility.

Lastly, gerontologiphobics do not seem to mind research into prevention of Alzheimer's disease, strokes, cancer or arthritis, and do not lodge protests against the provision of flu shots, insulin, pacemakers or appendectomies to those who need them. Each of these areas of research, and each of these and similar interventions, has as its goal the postponement of death and disability. Each is laudable and is embraced as laudable by society and pundits alike. Gerontologiphobics see nothing wrong in accepting the pleasant consequences of medical progress for themselves, for their loved ones and, if someone else is paying for it, even for casual acquaintances. Research on ageing has the same goal as research into other modalities of treatment or disease prevention: preservation of health and postponement of illness and disability. Those who oppose ageing research through some vague sense that it might be unnatural have not thought the matter through carefully enough to see the inconsistency of their intuitions.

A chapter this brief must of necessity be merely a synopsis. A reader who wishes to learn why serious scientists feel the need to disavow insupportable claims hawked by celebrity immortalists should read Warner et al (2005) for a recent consensus viewpoint. A reader who wants to learn more about where ageing research has gone and may soon be going can have a look at a hypothetical set of future landmarks (Miller, 1997), a textbook summary of the field's current status (Miller, 2003), or either of two books written for literate non-specialists (Austad, 1997; Kirkwood, 1999). And a reader who wants a somewhat more detailed presentation of the political and social obstacles impeding progress in biogerontology may wish to consult Miller (2002).

Part V

Stronger?

14

Engineering Challenges to Regenerative Medicine

Z. F. Cui

INTRODUCTION

Regenerative medicine will certainly make people 'stronger'. Its unique approach is to regenerate cells, tissues and even organs *in situ*, to prevent and cure diseases and to repair, restore and enhance functions of damaged and diseased tissues and organs. Common approaches in regenerative medicine include stem cell transplantation, cell therapy, gene therapy and tissue engineering, and combinations thereof. This chapter will only focus on the two relatively new and often related methods – stem cell-based therapy and tissue engineering, and later the engineering challenges for their commercialization and clinical applications.

TISSUE ENGINEERING

Tissue engineering was established as an identified field in the late 1980s, although it has to be pointed out that such an approach may have been being practised clinically for centuries. Its basic concept is to produce live and functional tissue by culturing appropriate cells on a three-dimensional scaffold for a certain period of time, and then to implant the engineered tissue into the body to replace or repair tissue that has been damaged or lost. The concept has been proved, and engineered skin products are available commercially.

However, apart from skin, which is avascular and of thin-layered structure, there has been no major breakthrough, either clinically or commercially, regarding any other tissues, despite continuous and huge effort into research and development of engineered bone, cartilage, small diameter blood vessels, pancreas, liver and so on. The reasons for this include:

1   Lack of understanding of biological systems at all levels, from the sub-cellular, through the cellular, tissue and organ, to the whole-body level. Important fundamental questions remain unanswered, such as:
    • what cells to use (for example differentiated cells vs. stem cells or banked cells vs. the patient's own cells);
    • how to culture them (medium, physical, chemical and mechanical conditions, and growth factors);
    • how to prevent host rejection; and
    • how to integrate the engineered tissue with the host.
2   Technical challenges and lack of an engineering approach. There have not been established engineering descriptions or even 'rules of thumb' for the design and operation of tissue engineering processes. Many laboratory tests followed the 'trial-and-error' approach, which, not surprisingly, had a low success rate.

These technical and engineering challenges cannot be underestimated, particularly considering clinical applications and commercialization. These challenges are described in the following sections.

Scaffold materials

Most materials tested in research and development were medically approved biomaterials, designed and approved for other applications such as sutures, wound dressing or implants. It is tempting to use these biomaterials for scaffold materials to speed up the regulatory approval, but they are not designed for tissue engineering scaffolds, and hence the chance of success is greatly reduced. Purpose-designed biomaterials, specific to engineered tissue types, will need to be developed to fulfil this requirement.

Bioreactor technology

The engineered tissue needs to be cultured in a bioreactor in a well-defined and well-controlled environment, including microenvironments around cells at different locations. Important parameters include nutrient and oxygen supply, metabolic waste removal, and hydrodynamic and mechanical stimuli. Designing such a bioreactor is not an easy task. Also, tissue culture takes a certain period of time, typically weeks, and is a costly process. Monitoring the process and adequate control is essential. Unfortunately, monitoring cell status within a three-dimensional tissue remains an unsolved problem, and hence a sensible control strategy cannot be implemented.

Bioprocess design

How to design a process to manufacture 1000 pieces of live tissue per week? Nobody has yet come up with a well-thought-out design methodology. The process must be scalable, economical and, more important, compliant with good manufacturer practice (GMP).

Preservation and storage of the engineered tissue products

Off-the-shelf availability is essential for commercial viability, although patient-specific, hospital-based tissue engineering is the low-hanging fruit. Cryopreservation is the process of choice, but it brings with it the great technical challenge of minimizing freezing damage to the product, and could increase the distribution costs significantly.

STEM CELL-BASED THERAPY

Stem cells offer unprecedented opportunities and potentials for regenerative medicine. Bone marrow transplantation is an established clinical procedure and the transplant of blood stem cells separated from umbilical cord blood has also been established. It is, however, the potential of using embryonic and adult stem cells to regenerate all the tissues and organs in the body that has attracted attention and imagination all over the world. If stem cell potentials are materialized, or even a fraction of their potentials are materialized, tomorrow's people will certainly live stronger and longer.

The ethical and social issues surrounding stem cell research and applications are well known, including where the stem cells come from and who will pay for the treatment. It must also be emphasized, however, that stem cell research and application is still in, if the reader will excuse the pun, an embryonic stage. There are many huge obstacles to be overcome before clinical application becomes a reality. Biological challenges include how to identify stem cells, how to culture stem cells, how to control their differentiation and how to follow their fate after implantation. Engineering challenges must also be addressed before clinical and commercial success can become a reality. Some of these issues are described in the following sections.

How to expand stem cells

Stem cells are rare and in very limited supply. For any therapy, a certain number of stem cells are required. Thus, a key issue is how to expand stem cells efficiently (i.e. grow and multiply by a large number) and in a controlled manner. This is more important for creating stem cell lines and stem cell banks. Using passages in cell culture dishes cannot meet this need. Bioreactor technology – using engineering methods to design and fabricate a well-defined environment for a large number of stem cells to grow and multiply – is essential.

How to differentiate stem cells

Controlled differentiation of stem cells to the desired cell type is the final step of tissue regeneration. Firstly, one needs to know exactly the quantitative relationship between cell local microenvironment and cell proliferation and differentiation. New research tools are required to assist in determining such links. BioMEMS and lab-on-chip engineering approaches could provide the required new tools. Secondly, how to achieve and deliver the required microenvironmental conditions to the local stem cells is a key issue. For *in vitro* tissue culture using stem cells, bioreactor technology again is the key, coupled with scaffold design. For *in vivo* cell therapy for tissue regeneration, design and selection of the materials injected together with the cells may hold the key; this may have to be based on a detailed mathematical modelling of the local environment.

How to preserve stem cells

Stem cells are precious. Thus stem cell storage and banking are essential steps for their clinical and commercial applications. Cryopreservation – storing the cells in liquid nitrogen – is the obvious choice. However, cryopreservation, even with the best available protecting chemicals, could cause the death of a significant proportion of cells. This is particularly true for human embryonic stem cells. Furthermore, when only some cells survive the procedure, cryopreservation acts like a selection step – in other words only the fittest survive, and these may not be the ones required. What is more, it is not necessarily the case that the stem cells that survive will still have all the original functions and potency. Hence there is an urgent need for new and efficient protocols for stem cell preservation to ensure maximum viability and functionality. In parallel, non-invasive, non-destructive techniques for testing and monitoring the quality of stem cells in storage need to be developed. If the stem cells are already dead, there is no point to keep them there for another 20 or more years.

CONCLUDING REMARKS

Tissue engineering and stem cell technology have great potential and will certainly make tomorrow's people live longer and stronger. However, materializing the promised potential is challenging and needs great and collected effort from clinicians, life scientists, physical scientists and engineers. An integrated, multi-disciplinary approach is essential, and alongside government and charity funding, there needs to be public support and understanding. A final remark to patients and investors – please be patient, the future is brighter.

15

Longevity and Regeneration

Ellen Heber-Katz

The possibility of human beings living longer implies processes of extended organ and tissue renewal. Due to various factors, ageing exposes the organism to tissue breakdown and replacement with less functional tissue, often due to micro-scar formation. This ongoing breakdown and repair could be described as a type of wound healing that is seen in mammals where one encounters a traditional injury with wound closure and extensive scarring. Virtually the entire literature on wound healing is devoted to this aspect of the healing process.

The alternative pathway is regeneration. This is most spectacularly seen in lower species such as newts, tadpoles, axolotls, zebrafish, sea cucumbers, hydra and planaria, where major injuries and limb, tail and body resections lead to a process of tissue and organ replacement through regrowth that is anatomically and functionally identical to the injured or lost organ. This is known as epimorphic regeneration (Goss, 1969). Besides the growth of new organs after injury, there also seems to be a high cell turnover rate in these species, providing a continual stream of new cells that are produced and lost at a steady rate and making for an organism that is continuously renewing itself, as seen in studies of hydra and planaria (David and Campbell, 1972; Otto and Campbell, 1977; Bode, 1996; Martinez, 1998; Newmark and Sanchez Alvarado, 2000; Gonzalez-Estevez and Salo, 2001; Galliot and Schmid, 2002; Holstein et al, 2003). However, the cellular/molecular relationship of these two phenomena remains to be determined.

The epimorphic regenerative response rarely happens in mammals, the few exceptions including antler regrowth in deer (Goss, 1969; Price et al, 2005). Ear hole closure in rabbits and wing hole closure in bats have also been included in this phenomenon (Goss and Grimes, 1975). In human beings, it is anecdotally noted that finger tips sometimes regrow (Douglas 1972; Illingworth, 1974) and studies in neonatal and foetal mice have shown that only injury through the nailbed

– the connective tissue underneath the nail – leads to regrowth (Borgens, 1982; Reginelli et al, 1995; Han et al, 2003). Thus, due to a paucity of actual examples, regeneration in mammals is largely unexplored territory. Furthermore, of these few cases, none provides a reproducible, genetically defined regeneration model in any species, let alone a biomedically important species such as the mouse.

The theme of regeneration in lower species is the presence of mechanisms that allow for the renewing of adult tissue that includes the de-differentiation of fully mature tissue to an immature state where cells can proliferate and then re-differentiate into different populations of cells of different types (Odelberg et al, 2000; Echeverri et al, 2001; Brockes and Kumar, 2005). For example, muscle fibres can eventually become chondrocytes.

There are alternative routes to regeneration, however. The participation of muscle stem cells or satellite cells has recently been reported in the amphibian newt (Morrison et al, 2006). In mammalian adults, participation of stem cells in tissue regeneration is seen in examples such as regrowth and regeneration of liver, epidermis, bone marrow and the gut lining. Furthermore, in recent years, the isolation of stem cell populations has provided a demonstrable source of cells capable of regenerating new tissue. Yet we still have to deal with the question of why mammals do not in fact regenerate. Even in the oft-cited case of liver regeneration, tissue does not grow back as well as is generally believed. This lack of robust regeneration in any of the higher mammals has naturally fostered the belief that regeneration had been lost through evolution. Perhaps there was a trade-off which substituted healing via scarring. In any event, regeneration is just beginning to be within the expectations of what medicine can hope to address.

Almost ten years ago, our laboratory accidentally discovered that the MRL mouse strain is capable of profound partial or complete regrowth in every organ system tested. This includes ear, digit and tail, heart, diaphragm, and nervous system (Heber-Katz et al, 2004a, and unpublished data). One of the first things we noticed about this MRL mouse was that there was a paucity of scar tissue after healing. The ear hole disappeared without a trace with no evidence of a scar. In fact, examination of liver healing in these mice showed a striking lack of adhesions after abdominal surgery. Was it possible that regeneration has been preserved through evolution? Did the regrowth seen in MRL mice parallel classical regeneration in newts? Several observations lead us to believe that this was the case.

One of the key events in amphibian limb regeneration after injury is the formation of the blastema, a structure of 'undifferentiated' cells which grows to a proper length and differentiates into functional digits, muscle, nerve and bone. Various patterning genes are expressed during this period (Carlson et al, 2001; Khan et al, 2002). Nothing similar to a blastema is found in adult mammals. They heal by the covering over of the wound with a thick scar and without the formation of new structures. That is why ear holes don't close and serve as a life-long marker in mice. However, examination of MRL ear hole sections during closure revealed a blastema that looked remarkably similar to the regenerating newt limb. Within a few months, the ear holes completely close with normal architecture, including cartilage, and without scarring.

Multiple markers have been identified in the MRL blastema that are shared with the amphibian; others are unique to the MRL. First, the inappropriate expression of keratin (generally uniquely expressed by epithelial cells) in the fibroblast cells of the newt blastema have been considered defining (Tassava et al, 1986; Ferretti et al, 1989). We have found this in the MRL blastema as well (Heber-Katz, 1999). Two other molecules, tenascin C and tbx5, have been shown to be upregulated in the newt and the MRL blastema (Tassava et al, 1996; Heber-Katz, 1999; Khan et al, 2002; Reing et al, in press). Molecules found only in the MRL blastema at this time include thrombospondin 1, thy 1 (CD90) and pref 1 (dlk) (Samulewicz et al, 2002; Reing et al, in press). These are only some of the defining molecules that are important in this phenomenon. But what makes the blastema grow, what makes it stop growing and what are the events that occur in between? We are already some way towards answering these questions.

One clue as to why there is a difference between mammals and amphibians came from an experiment done first by David Stocum (Stocum and Crawford, 1987) and suggested by previous findings (Stocum and Dearlove, 1972; Repesh and Oberpritter, 1980; Globus et al, 1980). Stocum later suggested this experiment in the MRL mouse. It is known that amphibians heal/regenerate by epithelial covering of the wound in the absence of a basement membrane. Stocum showed that induction of that basement membrane between the epidermis and dermis leads to a shutdown of the regenerative response and subsequent scarring. Examination of the MRL ear hole showed that the MRL starts the healing process like a wound repair event, with a provisional matrix and a basement membrane. However, soon after the epidermis covers the wound, the MRL breaks down that basement membrane and now allows a potential cross-talk between the dermis and epidermis. It is at this point that the blastema begins to grow. The remodelling response contributed to by at least the matrix metalloproteinases MMP-2 and MMP-9 is more pronounced in the MRL and includes enhanced inflammation with infiltrating cells that carry the MMPs (Gourevitch et al, 2002). This protease response has been shown to be important in the regenerative response in many non-mammalian systems (Grillo et al, 1968; Yang and Bryant, 1994; Miyazaki et al, 1996; Chernoff et al, 2000; Quinones et al, 2002) and may also relate to the lack of scar formation seen in the MRL compared to the normal mammalian scar result.

In fact, scarring may be the major reason why mammals do not regenerate. It is also a major problem in many diseases and is very prominent in the ageing process. We addressed this issue in a series of experiments (Seitz et al, 2002), employing a spinal injury model in which we confirmed all of the previous studies showing that a complete transection of the spinal cord led to loss of mobility and no evidence of recovery. In this case, the injury site was filled with an acellular scar and axonal and glial cells could not enter the injury site. However, in a unique type of injury made with a small incision in the dura covering the spinal cord far from the transection site and then a complete cord transection within the dura, we found that after three weeks, cord bridging occurred and axonal growth through the injury site was seen through these bridges. These animals showed labelling in the brain after a fluorescent dye was injected below the injury site and also recovered mobility. The most striking thing about this experiment was that it was done in the C57BL/6 mouse, our non-healer control. Though there are many caveats, this clearly showed that the normal mouse is capable of at least one form of regenerative response if scarring could be minimized.

Very early in our studies of MRL ear hole closure, we began microsatellite mapping to identify genetic loci, the strength of the inbred mouse model (McBrearty et al, 1998). Our initial studies showed 7 loci, and this has been expanded by several independent groups to over 20 loci at the current time, indicating a highly complex trait (Masinde et al, 2001; Blankenthorn et al, 2003; Heber-Katz et al, 2004b; Yu et al, 2005).

We next decided to look at the healing response of the diaphragm following a cryo-injury. It was immediately clear that the right ventricle of the heart was adjacent to the diaphragm and was easily accessible for cryo-injury. This heart wound model already existed in the literature (Taylor et al, 1998). The cryo-injury procedure was straightforward and reproducible, and the results were dramatic. Within 60 days following injury, almost complete replacement of myocardium with normal structure occurred (Leferovich et al, 2001). A central dogma of cardiobiology appeared to be overthrown. Mammalian hearts could regenerate without intervention. At the same time, studies appeared from various laboratories showing regenerative cardiac responses and stem cell transdifferentiation to cardiomyocytes in human beings (Ferrari et al, 1998; Beltrami et al, 2003; Lanza et al, 2004).

The heart cryo-injury leads to the death of cardiomyocytes and replacement by a fibroblast-like population. Within 30 days, there are islands of new cardiomyocytes, by 60 days the injury has been replaced by new cardiomyocytes and is histologically normal, and by day 90 echocardiography shows a normal functional response. We found that the cardiomyocytes that had filled the injury site were showing evidence of cell division. This is in contrast to what is seen in two other strains of mice tested, the C57BL/6 mouse and the Swiss Webster mouse, whose injuries end with an acellular scar that never heals. Of course, virtually all other mammals have cardiac injuries that end this way.

Further studies in the MRL mouse showed that the ability to heal heart tissue could be transferred with foetal liver from this mouse into non-healing mice. After injury, new cardiac tissue was seen. A second striking finding was that foetal liver cells from non-healer mice transferred scar formation into the MRL mouse after cardiac injury (Bedelbaeva et al, 2004). Recently we have more narrowly defined the populations that transfer these characteristics. This implies that 'stem cells' from individuals that scar and do not regenerate have cells that will transfer scarring. This has been noted recently in studies showing that recruitment of stem cells leads to cardiac scar formation (Zohlnhöfer et al, 2006; Kloner, 2006).

The anti-scarring protease response, so critical to ear hole regeneration, seems to be at work in the heart model as well. We found that a breakdown of collagen (the scar component) occurs over the full 60 days of healing in the MRL, whereas a collagen build-up occurs in the C57BL/6 mouse. The MMP responses are elevated in the MRL over the period of time of collagen breakdown and are poorly represented in the C57BL/6 mouse. Also, blocking the MMP response using the antibiotic minocycline in the MRL leads to poor heart regeneration and poor ear hole closure as well, further indicating the importance of collagen breakdown (Heber-Katz, unpublished data). A possible side-effect of this, however, is that minocycline treatment used for infections could result in scar formation when tissue is remodelled after infection and possibly block heart healing in the context of a simultaneously occurring myocardial infarction.

The parallels between regeneration in mammals and normal growth and development in foetal life are often noted. Does regeneration recapitulate development? Parsimony might suggest this, but parsimony arguments have often failed in biology. We decided to look at an accessible yet key element in the regenerative regulation that occurs between the tadpole and the adult stages of the frog. Tadpoles regenerate; adult frogs do not. The central factor is thyroid hormone. It has long been known that when thyroid hormone levels rise, regeneration ceases and metamorphosis begins (Berry et al, 1998). Injection of thyroid hormone into tadpoles blocks regeneration and induces metamorphosis. Does this in any way apply to mammals? Our answer is yes: when we injected MRL mice with T3 and T4 thyroid hormone, regeneration was blocked; when we injected propylthiouracil (PTU) to block thyroid hormone in C57BL/6 *non-healer* mice, regenerative responses were seen. This was specific to the heart (Heber-Katz, 2005). This confirms for us that there are deep connections between regeneration and development. Note that PTU is an approved drug for human use for the treatment of hyperthyroid disease. It is inexpensive and has been around for decades. Transient thyroid hormone shutdown could potentially lead to enhanced heart regeneration.

In the final analysis, how do we utilize these findings in terms of ageing? We can examine the problem as involving several issues. First, we need to promote cell proliferation so that organ regeneration can proceed. Second, we need to stop the scar response. We have shown that avoidance of a scar can lead to striking spinal cord recovery and is critical in ear hole and heart regeneration. The MRL avoids scar formation by at least one mechanism involving matrix metalloproteinases. A search for drugs which enhance the protease response could therefore be useful. We have also shown that reduced thyroid function may lead to enhanced heart regeneration and that thyroid replacement may lead to lowered regenerative capacity and enhanced scar formation. Finally, identification of the genes involved will tell us a lot about the processes taking place and allow a pharmacological approach to enhanced regeneration, reduced scarring and longer lives.

16

Augmenting Human Beings

Kevin Warwick

INTRODUCTION

In its fundamental form, the term 'cyborg' refers to a cybernetic organism, something that is part human, part machine. However, this can take on several guises. Some would perhaps regard all human beings as cyborgs in that we ride bicycles, drive cars, wear glasses, use computers and so on – essentially we are cyborgs because we interact with the technology around us (Clark, 2002). Then there is another grouping which includes those people that have been implanted with technology in order to provide a remedy to an ailment, possibly a heart pacemaker, a cochlear implant or even an artificial hip. This group arguably have a stronger claim in that they interact with and in many cases are dependent on their technological elements. But nevertheless their technological part is not allowing them enhanced capabilities but rather it is replacing ineffective human components.

In the world of science fiction the term cyborg has been used primarily to refer to a group in which a biological entity or elements and technology are inextricably linked to enhance the whole. One example would be the Six Million Dollar Man, who, having started life as a normal human being, has acquired extra powerful legs and super-senses due to his new oneness with technology. It is this form of cyborg in which we are most interested in this chapter – where a human being's abilities are enhanced by means of integral technology. But here the phenomenon will be looked at from a scientific viewpoint.

Research is being carried out (see, for example, Penny et al, 2000) in which biological signals are directly measured and are then used to either control a device or as an input to some type of feedback mechanism. In the majority of cases the signals are measured externally to the body, which presents communication and measurement problems (Wolpaw et al, 1991; Kubler et al, 1999). Whatever system is used, errors occur due to attenuation through the body and noise issues associated with small signal detection. When only external stimulation is apparent, for example mechanical vibration or thermal variation, then it is not possible to select unique sensory receptor channels, due to the general nature of the stimulation procedure.

Studies looking at the integration of technology with the human central nervous system have varied from diagnostic (Denislic and Meh, 1994), to the amelioration of symptoms (Popovic et al, 1998; Yu et al, 2001; Poboroniuc et al, 2002), to the augmentation of existing senses (Cohen et al, 1999; Butz et al, 1999). But the most widely reported research involving human subjects is that based on the development of an artificial retina (Kanda et al, 1999). In this work, small arrays have been successfully attached to a functioning optic nerve, but where the person has no vision. By means of direct stimulation of the nerve with appropriate signal sequences, this has resulted in the user perceiving simple shapes and even letters. We must be clear, though, that this approach cannot yet instantly restore even a limited form of sight.

Electronic neural stimulation has proved to be extremely successful in other areas, with applications ranging from cochlear implants to the treatment of Parkinson's disease symptoms. The example most relevant to this chapter is the use of a brain implant which enables a brainstem stroke victim to control the movement of a cursor on a computer screen (Kennedy et al, 2000 and 2004). A hollow glass electrode cone containing two gold wires was implanted into the motor cortex such that when the patient thought about moving his hand the output from the electrode was amplified and transmitted by a radio link to a computer, where the signals were translated into control signals to bring about movement of the cursor.

In each of the cases in which human subjects have been involved, the aim has been either to bring about some restorative functions when an individual has a physical problem of some kind, for example blindness, or to give a new ability to an individual who has very limited abilities of any kind due to a major malfunction in their brain or nervous system. Here I am concerned with neither of these situations. Rather I wish to consider the possibility of giving extra capabilities to an individual human being.

THE RFID IMPLANT

In 1997 various scientists were pointing to a future in which we would no longer need passports and keys – rather all we would have would be a small piece of silicon under the skin (Cochrane, 1997). So on 24 August 1998 I became the first human being to have a Radio Frequency Identification Device (RFID) surgically implanted in my left arm. With this in place, when I walked around the cybernetics building at Reading University, the computer could track my movements. It said hello to me when I came through the front door, switched on lights for me in the corridor and opened the laboratory door automatically when I approached (Warwick, 2002). What it didn't do, though, was to increase my senses in any way or even interact inside my body. Clearly something more was required for that.

THE MICROELECTRODE ARRAY

On 14 March 2002, during a two-hour procedure at the Radcliffe Infirmary, Oxford, a microelectrode array was surgically implanted into the median nerve fibres of my left arm. The array measured 4×4mm, with each of the electrodes being 1.5mm in length. With the median nerve fascicle estimated to be 4mm in diameter, the electrodes penetrated well into the fascicle. The array was pneumatically inserted into the median nerve fibres, such that the body of the array sat adjacent to the fibres, with the electrodes penetrating into the fascicle.

The array was positioned just below the wrist, following a 4cm-long incision. A further incision, 2cm long, was made 16cm proximal to the wrist. The two incisions were connected by a tunnelling procedure such that wires from the array ran up

Figure 16.1 *A 100 electrode, 4×4mm MicroElectrode Array, shown on a UK 1 pence piece for scale*

the inside of the left arm, where they exited and were connected to an electrical terminal pad which remained external.

The arrangements described above remained in place for 96 days until 18 June 2002, when the implant was successfully removed.

NERVE STIMULATION

The array, once in position, acted as a neural interface. Signals could be transmitted directly from my nervous system to a computer and also from a computer to the array to directly bring about a stimulation of my nervous system. A technical description of the experiment can be found in a number of papers and hence will not be repeated here (Warwick et al, 2003 and 2004; Gasson et al, 2005).

Stimulation of the nervous system by means of the array was especially problematic due to the extremely limited nature of existing results using this type of interface. Previously published work was restricted largely to a respectably thorough but short-term study into the stimulation of the sciatic nerve in cats. Much experimental time was therefore required, on a trial and error basis, to ascertain what voltage/current relationships would produce a reasonable (in other words perceivable but not painful) level of nerve stimulation.

Further factors which may well turn out to be relevant, but were not possible to predict before the experimental session were:

- the plastic, adaptable nature of the human nervous system and the brain – even over relatively short periods; and
- the effects of movement of the array in relation to the nerve fibres – hence the connection and associated input impedance of the nervous system were not completely stable.

It should also be noted that it took six weeks for my brain to repetitively recognize the stimulating signals accurately. This time period may be due to a number of contributing factors:

- the team had to learn which signals (what amplitude, frequency and so on) would be best in order to bring about a recognizable stimulation;
- my brain had to learn to recognize the new signals it was receiving; and
- the bond between my nervous system and the implant was physically changing (becoming stronger).

However, it was found to be possible to create alternative sensations via this new input route to the nervous system.

AUGMENTED SENSORY INPUT

A further experiment was set up to determine if the human brain is able to understand and successfully operate with sensory information to which it had not previously been exposed. Whilst it is quite possible to feed in such sensory information via a normal human sensory route, for example when electromagnetic radar or infrared signals are converted to visual as with night sights, what we were interested in was feeding such signals directly into the human nervous system, thereby bypassing the normal human sensory input.

Ultrasonic sensors were fitted to the rim of a baseball cap (see Figure 16.2) and the output from these sensors, in the form of a proportional count, was employed to bring about a direct stimulation of my nervous system. When no objects were in the vicinity of the sensors, no stimulation occurred, and as an object moved close by, so the rate of stimulation pulses being applied increased in a linear fashion up to a preselected maximum rate. No increase in stimulation occurred when an object moved closer than 30cm to the sensors.

It was found that very little learning was required for the new ultrasonic sense to be used effectively and successfully – merely a matter of 5–6 minutes. This said, it must be remembered that it had already taken several weeks for my brain to successfully, accurately recognize the current signals being injected.

Figure 16.2 *Experimentation and testing of the ultrasonic baseball cap*

As a result, in a witnessed experiment, whilst wearing a blindfold, I was able to move around successfully within a cluttered laboratory environment, albeit at a slower than normal walking pace. The sensory input was 'felt' as a new form of sensory input (not as touch or movement) in the sense that the brain made a direct link between the signals being witnessed and the fact that these corresponded in a linear fashion to a nearby object.

Meanwhile, in another experiment my nervous system was linked directly to the internet and I was able to not only directly drive an articulated hand by means of my neural signals but also to feel the force the fingertips of the hand were applying to objects. In this way, as a cyborg, my nervous system was extended over the internet.

Discussion

One aspect of this study is the potential use of these types of implants to help those with a lesion in their nervous system, the aim being to bring about some otherwise missing movement; to return control of body functions to the body's owner (Finn and LoPresti, 2003); or to allow a recipient to operate technology around them (Leuthardt et al, 2004), generally with the target of servicing the communication and control needs of people with severe motor disabilities.

The particular line of research described in this chapter could also have an immediate impact for those who are blind, not by repairing their blindness, but by giving them an alternative sense. This would allow them to move around in the world with an ultrasonic knowledge of any nearby objects. Certainly the success of this one-off study indicates that it may be possible for a person who is blind to be fitted in the same way with an implant and make relatively rapid use of it. A big question remains, however, as to how different brains adapt to such a new input stream.

A further aspect of the research was to investigate the human body's acceptance or rejection of such an implant. No infection whatsoever was witnessed during the course of the array experiment, and during extraction it was observed that body tissue had grown around the array, holding it in its original place.

I am aware that this chapter only describes a one-off pilot study based on only one recipient – myself. It may well be that other recipients react in other ways. And while the study can be taken as practical evidence of the concept, before any large-scale conclusions can be drawn on the usefulness and effect of such sensory input, many other trials will be necessary. My team are now involved in such studies, and further implant trials will occur in due course, as indeed will implant studies directly linking together the human brain and computers.

Part VI

Smarter?

17

Brain Boosters

Nick Bostrom and Anders Sandberg

INTRODUCTION

Cognitive enhancement may be defined as the amplification or extension of core capacities of the mind through improvement or augmentation of internal or external information processing systems. The spectrum of cognitive enhancements includes not only medical interventions, but also psychological interventions (such as learned 'tricks' or mental strategies), as well as improvements of external technological and institutional structures that support cognition. A distinguishing feature of cognitive enhancements, however, is that they improve *core cognitive capacities* rather than merely particular skills or domain-specific knowledge.

Most efforts to enhance cognition are of a mundane nature, and some have been practised for thousands of years. The prime example is education and training, where the goal is often not only to impart specific skills or information, but also to improve general mental faculties such as concentration, memory and critical thinking. Other forms of mental training, such as yoga, martial arts, meditation and creativity courses, are also in common use. Caffeine is widely used to improve alertness. Herbal extracts reputed to improve memory are popular, with sales of *Ginko biloba* alone in the order of several hundred million dollars per year in the US (van Beek, 2002). In an ordinary supermarket, we find a staggering number of energy drinks on display, vying for consumers who are hoping to turbo-charge their brains.

Education and training, as well as the use of external information processing devices, may be labelled as 'conventional' means of enhancing cognition. They are often well established and widely culturally accepted. By contrast, methods of enhancing cognition through 'unconventional' means, such as ones involving deliberately created nootropic drugs, gene therapy or neural implants, are nearly all regarded as experimental at the present time. Nevertheless, these unconventional forms of enhancements are worthy of serious consideration, not only because of their novelty but also because they might eventually offer enormous leverage – consider the cost/benefit ratio of a cheap pill that safely enhances cognition compared to years of extra education.

EDUCATION, ENRICHED ENVIRONMENTS AND GENERAL HEALTH

Education has many benefits beyond higher job status and salary. Longer education reduces the risks of substance abuse, crime and many illnesses while improving quality of life, social connectedness and political participation (Johnston, 2004). There is also positive feedback between performance on cognitive tests such as IQ tests and scholastic achievement (Winship and Korenman, 1997).

Much of what we learn in school is 'mental software' for managing various cognitive domains: mathematics, categories of concepts, language and problem-solving in particular subjects. This kind of mental software reduces our mental load through clever encoding, organization or processing. Specialized methods have a smaller range of applicability but can dramatically improve performance within a particular domain. They represent a form of crystallized intelligence, distinct from the fluid intelligence of general cognitive abilities and problem-solving capacity (Cattell, 1987). The relative ease and utility of improving crystallized intelligence and specific abilities have made them popular targets of internal and external software development. Cognitive enhancement attempts the more difficult challenge of improving fluid intelligence.

Pharmacological cognitive enhancements (nootropics) have physiological effects on the brain. So too do education and other conventional interventions. In fact, conventional interventions often produce more permanent neurological changes than do drugs. Learning to read alters the way language is processed in the brain (Petersson et al, 2000). Enriched rearing environments have been found to increase dendritic arborization and to produce synaptic changes, neurogenesis and improved cognition in animals (Walsh et al, 1969; Greenough and Volkmar, 1973; Diamond et al, 1975; Nilsson et al, 1999). Enriched environments also make brains more resilient to stress and neurotoxins (Schneider et al, 2001). Reducing neurotoxins and preventing bad prenatal environments are simple and widely accepted methods of increasing cognitive functioning. These latter kinds of intervention might be classified as preventative or therapeutic rather than enhancing, although the distinction is blurry. For instance, an optimized intrauterine environment will not only help avoid specific pathology and deficits, but is also likely to promote the growth of the developing nervous system in ways that ultimately *enhance* its core capacities.

Improving general health has cognition-enhancing effects. Many health problems act as distractors or directly impair cognition (Schillerstrom et al, 2005). Improving sleep, immune function and general conditioning promotes cognitive functioning. Bouts of exercise have been shown to temporarily improve various cognitive capacities (Tomporowski, 2003). Long-term exercise also improves cognition, possibly by a combination of increased blood supply to the brain and the release of nerve growth factors (Vaynman and Gomez-Pinilla, 2005).

MENTAL TRAINING

Mental training and visualization techniques are widely practised in elite sport (Feltz and Landers, 1983) and rehabilitation (Jackson et al, 2004), with apparently good effects.

Even general mental activity, 'working the brain muscle', can improve performance (Nyberg et al, 2003) and long-term health (Barnes et al, 2004), as can relaxation techniques to regulate the activation of the brain (Nava et al, 2004). It has been suggested that the Flynn effect (Flynn, 1987), a secular increase in raw intelligence test scores by 2.5 IQ points per decade in most Western countries, can be attributed to increased demands of certain forms of abstract and visuo-spatial cognition in modern society and schooling, although improved nutrition and health status may also play a part (Neisser, 1997; Blair et al, 2005). It appears that most of the Flynn effect does not reflect an increase in general fluid intelligence but rather a change in which specific forms of intelligence are developed.

The classic form of cognitive enhancement software is learned strategies to memorize information. Such methods have been used since antiquity with much success (Yates, 1966; Patten, 1990). In general it appears possible to attain very high memory performance on specific types of material using memory techniques. These work best on otherwise meaningless or unrelated information such as sequences of numbers, but do not appear to help skilled everyday activities (Ericsson, 2003).

There also exists a vast array of mental techniques alleged to boost various skills, such as creativity training, speed reading methods (Calef et al, 1999) and mind maps (Buzan, 1982; Farrand et al, 2002). Good data regarding their efficacy is often lacking, however. Even if a technique improves performance on some task under particular conditions, that does not necessarily mean that the technique is practically useful. In order for a technique to significantly benefit someone, it would have to be effectively integrated into his or her everyday work.

DRUGS

Stimulant drugs like nicotine and caffeine are traditionally used to improve cognition. In the case of nicotine, a complex interaction with attention and memory occurs (Warburton, 1992; Newhouse et al, 2004; Rusted et al, 2005), while caffeine reduces tiredness (Lieberman, 2001; Smith et al, 2003; Tieges et al, 2004). Today there exist a broad range of drugs that can affect cognition (Farah et al, 2004).

Lashley observed in 1917 that strychnine facilitates learning in rats (Lashley, 1917). Since then several families of memory-enhancing drugs affecting different aspects of long-term memory have been discovered. These range from stimulants (Soetens et al, 1993 and 1995; Lee and Ma, 1995), nutrients (Korol and Gold, 1998; Winder and Borrill, 1998; Meikle et al, 2005) and hormones (Gulpinar and Yegen, 2004), through cholinergic agonists (Iversen, 1998; Power et al, 2003; Freo et al, 2005) and the piracetam family (Mondadori, 1996), to ampakines (Ingvar et al, 1997; Lynch, 1998) and consolidation enhancers (Lynch, 2002). Glucose is the major energy source for the brain, which relies on a continuous supply to function, and increases in glucose availability (for example due to ingestion of sugars or the release of the acute stress hormone noradrenalin) improve memory (Wenk, 1989; Foster et al, 1998).

The earliest enhancer drugs were mainly non-specific stimulants and nutrients. During antiquity, honey water, hydromel, was used for doping purposes. But advances in our understanding of memory has allowed the development of more specific drugs. Stimulating the cholinergic system, which appears to gate attention and memory encoding, was a second step. Current interest is focused on intervening into the process of permanent encoding in the synapses, which has been elucidated to a great extent and hence has become a promising target for drug development. The goal would be drugs that not just allow the brain to learn quickly but also facilitate selective retention of the information that has been learned. It is known that the above families of drugs can improve performance in particular memory tests. It is not yet known whether they also promote useful learning in real-life situations.

Pharmacological agents might be useful not only for increasing memory retention, but also for unlearning phobias and addictions (Pitman et al, 2002; Ressler et al, 2004; Hofmann et al, 2006). Potentially, the combination of different pharmacological agents administered at different times could allow users a more fine-grained control of their learning processes, and perhaps even the ability to deliberately select the contents of their memory.

Even common, traditional and unregulated herbs and spices such as sage can improve memory and mood through chemical effects (Kennedy et al, 2006). While less powerful than those of dedicated cholinesterase inhibitors, such effects illustrate that attempts to control access to cognition-enhancing substances would be problematic. Even chewing gum appears to affect memory, possibly by heightening arousal or blood sugar (Wilkinson et al, 2002).

Working memory can be modulated by a variety of drugs. Drugs that stimulate the dopamine system have demonstrated effects, as do cholinergic drugs (possibly through improved encoding) (Barch, 2004). Modafinil has been shown to enhance working memory in healthy test subjects, especially at harder tasks and for lower-performing subjects (Muller et al, 2004). (Similar findings of greater improvements among low performers were also seen among the dopaminergic drugs, and this might be a general pattern for many cognitive enhancers.) On a larger battery of tasks, modafinil was found to increase forward and backward digit span, visual pattern recognition memory, spatial planning, and reaction time/latency on different working memory tasks (Turner et al, 2003a).

There also exist drugs that influence how the cerebral cortex reorganizes in response to damage or training. Noradrenergic agonists such as amphetamine have been shown to promote faster recovery of function after a brain lesion when combined with training (Gladstone and Black, 2000) and to improve learning of an artificial language (Breitenstein et al, 2004).

TRANSCRANIAL MAGNETIC STIMULATION

Transcranial magnetic stimulation (TMS) can increase or decrease the excitability of the cortex, thereby changing its level of plasticity (Hummel and Cohen, 2005). TMS of the motor cortex that increased its excitability improved performance in a procedural learning task (Pascual-Leone et al, 1999). TMS in suitable areas has also been found beneficial in a motor task (Butefisch et al, 2004), motor learning (Nitsche et al, 2003), visuo-motor coordination tasks (Antal et al, 2004a and 2004b), working memory (Fregni et al, 2005), finger sequence tapping (Kobayashi et al, 2004), classification (Kincses et al, 2004) and even declarative memory consolidation during sleep (Marshall et al, 2004). While TMS appears to be versatile and non-invasive, there are risks of triggering epileptic seizures, and the effects of long-term use are not known. Individual brain differences may necessitate much adjustment before specific application.

GENETIC MODIFICATIONS

Genetic memory enhancement has been demonstrated in rats and mice. Genetically modified 'Doogie' mice demonstrated improved memory, both acquisition and retention. The modification also made the mice more sensitive to certain forms of pain, suggesting a non-trivial trade-off between two potential enhancement goals (Wei et al, 2001).

Increased amounts of brain growth factors (Routtenberg et al, 2000) and the signal transduction protein adenylyl cyclase (Wang et al, 2004) have also produced memory improvements in animal models. The cellular machinery of memory appears to be highly conserved in evolution, suggesting that interventions demonstrated in animal models might also work in people (Edelhoff et al, 1995; Bailey et al, 1996).

Genetic studies have also found genes in human beings whose variations account for up to 5 per cent of memory performance (de Quervain and Papassotiropoulos, 2006). These include the genes for the NMDA receptor and adenylyl cyclase.

Given these early results, it seems likely that there exist many potential genetic interventions that might directly or indirectly improve aspects of memory. If it turns out that the beneficial effects of the treatments are not due to changes in development, then presumably some of the effects can be achieved by supplying the brain with the substances produced by the memory genes without resorting to genetic modification.

On the other hand, studies of the genetics of intelligence suggests that there are a large number of genetic variations affecting individual intelligence, but each accounting for only a very small fraction (<1 per cent) of the variance between individuals (Craig and Plomin, 2006). This would indicate that genetic enhancement of intelligence through direct insertion of a few beneficial alleles is unlikely to have a big enhancing effect. It is possible, however, that some alleles that are rare in the human population could have larger effects on intelligence, either negative or positive.[1]

PRENATAL AND PERINATAL ENHANCEMENT

Administering choline supplementation to pregnant rats improved the performance of their pups, apparently as a result of changes in neural development (Meck et al, 1988; Mellott et al, 2004). Given the ready availability of choline supplements, such prenatal enhancement may already (inadvertently) be taking place in human populations.

Supplementation of a mother's diet during late pregnancy and three months postpartum with long-chained fatty acids has been demonstrated to improve cognitive performance in human children (Helland et al, 2003). Deliberate changes of maternal diet may hence be seen as part of the cognitive enhancement spectrum. At present, recommendations to mothers are mostly aimed at promoting a diet that avoids specific harms and deficits, but the growing emphasis on boosting 'good fats' and the use of enriched infant formulas point towards enhancement.

EXTERNAL HARDWARE AND SOFTWARE SYSTEMS

Some approaches in human–computer interaction explicitly aim at cognitive enhancement (Engelbart, 1962). External hardware is of course already used for cognition enhancement, be it pen and paper or computer software like personal organizers. Many common pieces of software act as cognition enhancing environments where the software helps give an overview, keep multiple items in memory and perform routine tasks. Data-mining and information-visualization tools help produce overview and understanding where the perceptual system cannot handle the amount of data, while specialized tools like expert systems, symbolic maths programs, decision-support tools and search agents expand specific skills and capacities.

What is new is the growing interest in creating intimate links between external systems and the human user through better interaction. The software becomes less an external tool and more of a mediating 'exoself'. This can be achieved through mediation, embedding the person within an augmenting 'shell' such as wearable computers (Mann, 2001; Mann and Niedzviecki, 2001) or virtual reality, or through smart environments where capabilities of objects in the environment are extended. An example is the ubiquitous computing vision, in which objects would be equipped with unique identities and given ability to communicate with and to support the user (Weiser, 1991). A well-designed environment can enhance proactive memory (Sellen et al, 1996) by deliberately bringing previous intentions to mind in the right context.

Another form of memory-enhancing exoself software is remembrance agents (Rhodes and Starner, 1996), agents that act as a vastly extended associative memory. The agents have access to a database of previous information, such as a user's files, email correspondence and so on, and suggest relevant documents based on the current context. Other exoself applications include additions to vision (Mann, 1997), team coordination (Fan et al, 2005a and 2005b), face recognition (Singletary and Starner, 2000), mechanical prediction (Jebara et al, 1997) and recording emotionally significant events (Healey and Picard, 1998).

Given the availability of external memory support, from writing to wearable computers, it appears likely that the crucial form of memory demand will be the ability to link together information into usable concepts and associations rather than storage and retrieval of raw data. Storage and retrieval functions can be offloaded to a great extent from the brain, while the knowledge, strategies and associations linking the data to skilled cognition so far cannot generally be offloaded.

BRAIN–COMPUTER INTERFACES

Wearable computers and personal digital assistants (PDAs) are already intimate devices worn on the body, but there have been proposals for even tighter interfaces. Control of external devices through brain activity has been studied with some success for the last 40 years, although it remains a slow form of signalling (Wolpaw et al, 2000).

The most dramatic potential internal hardware enhancements are brain–computer interfaces. At present development is rapid both on the hardware side, where multi-electrode recordings from more than 300 electrodes permanently implanted in the brain are currently state-of-the-art, and on the software side, with computers learning to interpret signals and commands (Nicolelis et al, 2003;

Shenoy et al, 2003; Carmena et al, 2003). Early experiments on people have shown that it is possible for profoundly paralysed patients to control a computer cursor using just a single electrode implanted in the brain (Kennedy and Bakay, 1998), and experiments by Patil et al have demonstrated that the kind of recordings used in monkeys would probably function in people too (Patil et al, 2004). Experiments in localized chemical release from implanted chips also suggest the possibility to use neural growth factors to promote patterned local growth and interfacing (Peterman et al, 2004).

Cochlear implants are already widely used, and there is ongoing research in artificial retinas (Alteheld et al, 2004) and functional electric stimulation for paralysis treatment (von Wild et al, 2002). The digital parts of the implant can in principle be connected to almost any kind of software and external hardware. This would enable enhancing uses such as access to software help, the internet and virtual reality applications. Non-disabled people, however, would probably achieve the same benefits through eyes, finger and voice control.

COLLECTIVE INTELLIGENCE

Much of human cognition is distributed across many minds and can be enhanced by developing more efficient forms of collaboration. The World Wide Web and email are among the most powerful kinds of cognitive enhancement software developed to date. Through the use of such social software, the distributed intelligence of large groups can be shared and harnessed for particular purposes (Surowiecki, 2004).

Connected systems allow many people to collaborate in the construction of shared knowledge and solutions: the more individuals that connect, the more powerful the system becomes (Drexler, 1991). The information is not stored just in the documents themselves but in their interrelations. When such interconnected information resources exist, automated systems such as search engines (Kleinberg, 1999) can extract a wealth of useful information from them.

Lowered coordination costs enable larger groups to work on common projects. Such groups of shared interests, such as amateur journalist 'bloggers' and open-source programmers, have demonstrated that they can successfully complete large projects, such as online political campaigns, the Wikipedia encyclopaedia and the Linux operating system. Systems for online collaboration can incorporate efficient error-correction (Raymond, 2001; Giles, 2005), enabling incremental improvement of product quality over time.

An interesting variant of knowledge aggregation is prediction markets (also known as 'information markets' or 'idea futures markets'). Here participants trade in predictions of future events, and the levels of their bets tend to reflect the best information available on the probability of whether the events will occur (Hanson et al, 2003). Such markets appear to be self-correcting and resilient and have been shown to outperform alternative methods of generating probabilistic forecasts, such as opinion polls and expert panels (Hanson et al, 2006).

ASSESSMENT

'Conventional' means of cognitive enhancement (such as education, mental techniques, neurological health and external systems) are largely accepted, while 'unconventional' means (such as drugs, implants and direct brain–computer interfaces) tend to evoke more moral and legal concerns. However, the demarcation between the two is blurry. The newness of the unconventional means – and the fact that they are currently still mostly experimental – may be more responsible for their problematic status, rather than any essential problem in the technologies themselves. As we learn more about their strengths, weaknesses, potentials for abuse and serendipitous benefits, through use and experience, these currently unconventional technologies may become absorbed into the ordinary discourse of human tools.

At present, most biomedical enhancement techniques produce at most modest improvements of performance (about 10–20 per cent improvement in a typical test task). More dramatic results can be achieved using training and human–machine collaboration, techniques that are less ethically controversial at present. Mental techniques can achieve 1000 per cent or more improvement in narrow domains, such as specific memorization tasks (Ericsson et al, 1980).

While pharmacological cognitive enhancements do not produce dramatic improvements on specific tasks, their effects are often quite general, enhancing performance on, for example, all different tasks making use of working or long-term memory. External tools and cognitive techniques such as memorization, on the other hand, are usually task-specific, producing potentially huge improvements of narrow abilities. A combination of different methods can be expected to do better than the individual technologies, especially in everyday or workplace settings where a wide variety of tasks have to be done.

Even small improvements in general cognitive capacities can have important positive effects. Individual cognitive capacity (imperfectly estimated by IQ scores) is positively correlated with income. One study estimates the increase in income from one additional IQ point to 2.1 per cent for men and 3.6 per cent for women (Salkever, 1995). Higher cognitive abilities also appear to prevent a wide array of social and economic misfortunes (Gottfredson, 1997 and 2004) and promote health (Whalley and Deary, 2001). At a societal level, the sum of many individual enhancements may be even more profound: economic models of the loss caused by small intelligence decrements due to lead in drinking water predict significant effects of even a few points of change (Salkever, 1995; Muir and Zegarac, 2001).

POLICY IMPLICATIONS

Many extant forms of regulation are intended to protect and improve cognitive function. Regulation of lead in paint and tap water, requirements of boxing, bicycle and motorcycle helmets, bans on alcohol for minors, mandatory education, folic acid fortification of cereals, legal sanctions against mothers taking drugs during pregnancy: all these serve to safeguard or promote cognition. To a large extent, these efforts are a subset of general health protection measures, yet stronger efforts appear to be made when cognitive function is at risk. One might also observe that mandated information duties, such as labelling of food products, were introduced to give consumers access to more accurate information in order to enable them to make better choices. Given that sound decision-making requires both reliable information and the cognitive ability to retain, evaluate and use this information, one would expect that enhancements of cognition will also promote rational consumer choice. By contrast, we know of no public policy that is intended to limit or reduce cognitive capacity. Insofar as patterns of regulation reflect social preferences, then, it seems that society shows at least an implicit commitment to better cognition.

At the same time, however, there exist a number of obstacles to the development and use of cognitive enhancements. One important obstacle is the present system for licensing drugs and medical treatments. This system was created to deal with traditional medicine, which aims to prevent, detect, cure or mitigate diseases. In this framework, there is no room for enhancing medicine. For example, drugs companies could find it difficult to get regulatory approval for a pharmaceutical whose sole use was to improve cognitive functioning in the healthy population. To date, every pharmaceutical on the market that offers some potential cognitive enhancement effect was developed to treat some specific disease condition (such as ADHD, narcolepsy or Alzheimer's disease). The cognitive enhancing effects of these drugs in healthy subjects is a serendipitous unintended effect. Progress in this area would almost certainly be accelerated if pharmaceutical companies could focus directly on developing nootropics for use in non-diseased populations rather than having to work indirectly by demonstrating that the drugs are also efficacious in treating some recognized disease.

One of the perverse effects of the failure of the current medical framework to recognize the legitimacy and potential of enhancement medicine is the trend towards medicalization and 'pathologization' of an increasing range of conditions that were previous regarded as part of the normal human spectrum. If a significant fraction of the population could obtain certain benefits from drugs that improve, for example, concentration, it is currently necessary to categorize this segment of people as having some disease – in this case attention deficit hyperactivity disorder – in order to get the drug approved and prescribed to those who could benefit from it. This disease-focused medical model is increasingly inadequate for an era in which many people will be using medical treatments for enhancement purposes.

The medicine-as-treatment-for-disease framework creates problems not only for pharmaceutical companies, but also for users ('patients'), whose access to enhancers is often dependent on being able to find an open-minded physician who will prescribe the drug. This creates inequities in access. People with high social capital and good information get access while others are excluded.

Today we are seeing the growth of personalized medicine, both as a result of improved diagnostic methods that enable a better picture of the individual patient and because of the availability of a wider range of medical interventions that make possible selection of treatment for particular patient needs. Many patients now approach their physicians armed with detailed knowledge of their condition and possible treatments, which might have been obtained from the internet or other sources. These factors are leading to a shift in the physician–patient relationship, away from paternalism to a relationship characterized by teamwork and a focus on the customer's situation. Preventative and enhancing medicine are often inseparable, and both will probably be promoted by these changes and by the increasingly active and informed patient insisting on exercising medical choice. Again, these shifts suggest the need for important and complex regulatory change.

Given that all medical interventions carry some risk, and that the benefits of enhancements may often be more subjective and value-dependent than the benefits of being cured of a disease, it is important to allow individuals to determine their own preferences for trade-offs between risks and benefits. It is highly unlikely that one size will fit all. At the same time, many will feel that there should remain a role for a limited degree of paternalism, to protect individuals from at least the worst risks. One option would be to establish some baseline level of acceptable risk in allowable interventions, perhaps by comparison to other risks that society allows individuals to take, such as risks from smoking, mountain climbing or horseback riding. Enhancements that could be shown to be no more risky than these activities would be allowed (with appropriate information and warning labels where necessary). Another possibility would be enhancement licences. People willing to undergo potentially risky but rewarding enhancements could be required to demonstrate sufficient understanding of the risks and the ability to handle them responsibly. This would both ensure informed consent and enable better monitoring. (A downside with enhancement licences is that people with low cognitive capacity, who may have the most to gain, could find it hard to get access if the licence requirements were too demanding.)

Public funding for research does not yet reflect the potential personal and social benefits of many forms of cognitive enhancement. There is funding (albeit arguably at inadequate levels) for research into education methods and information technology, but not for pharmacological cognitive enhancers. In view of the potentially enormous gains from even moderately effective general cognitive enhancements, this area deserves large-scale funding. It is clear that much research and development are needed to make cognitive enhancement practical and efficient. As discussed above, this requires a change in the view that medicine is only about restoring, not enhancing, capacities and how this view is expressed in regulations of medical trials and drug approval.

The evidence on prenatal and perinatal nutrition suggests that infant formulas containing suitable nutrients may have a significant positive life-long impact on cognition. Because of the low cost and extremely large potential impact of enriched infant formula if applied at a population level, it should be a priority to conduct more research to establish the optimal composition of infant formula. Regulation could then be used to ensure that commercially available formula contains these nutrients. Public health information campaigns could further promote the use of enriched formula that promotes mental development. This would be a simple extension of current regulatory practice, but a potentially important one.

There is a wider cultural challenge in destigmatizing the use of enhancers. At present, the taking of medicine is regarded as a regrettable condition, and use of non-treatment medication is seen as suspect, or possibly misuse. Attempts to enhance cognition are often construed as the expression of a dangerous ambition. Yet the border between accepted theory and suspect enhancement is shifting. Plastic surgery enjoys ever-wider acceptance. Millions of people ingest nutrient supplements and herbal remedies for enhancing purposes. Self-help psychology is very popular. Apparently, the cultural constructions surrounding the means of enhancement are more important for their acceptance than the actual enhancement ability of these means. To make the best use of our new opportunities, therefore, we need a culture of enhancement, with norms, support structures and a lay understanding of enhancement that takes it into the mainstream cultural context. Consumers need better information on the risks and benefits of enhancers, which suggests a need for reliable consumer information and testing.

Testing of cognitive enhancers would ideally be done not only in the lab, but also in field studies that investigate how an intervention works in everyday life. The ultimate criterion of efficacy would be various forms of life success rather than performance in narrow psychological lab tests. Such 'ecological testing' would require new kinds of experiment, including monitoring of large sample populations. Advances in wearable computers and sensors may allow unobtrusive monitoring of behaviour, diet, use of other drugs and so on. Data-mining of collected materials could help determine the effects of enhancers. Such studies, however, pose major challenges, including cost, new kinds of privacy concerns (monitoring may accumulate information not only about the consenting test subjects but also about their friends and family), and problems of unfair competition if enhancers experience beneficial effects but others cannot get access to the enhancements due to their experimental nature.

While currently access to medicine is regarded as a human right constrained by cost concerns, it is less clear whether access to all enhancements should or would be regarded as a positive right. The case for at least a negative right to cognitive enhancement, however, based on cognitive liberty, privacy interests, and the important interest of people to protect and develop their own minds and capacity for autonomy, seems very strong.[2] Banning enhancements would create an incitement for black markets as well as limit socially beneficial uses. Legal enhancement would promote development and use, in the long run leading to cheaper and safer enhancements. Yet without public funding, some useful enhancements may be out of reach to many of the people who would benefit the most from them. Proponents of a positive right to enhancements could argue their position on grounds of fairness or equality or on grounds of a public interest in the promotion of the capacities required for autonomous agency. The societal benefits of effective cognitive enhancement may even turn out to be so large and clear that it would be Pareto optimal to subsidize enhancement for the poor, just as the state now subsidizes education.

Notes

1   A possible example is suggested in Cochran et al (2006), where it is predicted that heterozygoticity for Tay-Sachs disease should increase IQ by about five points.
2   For arguments for a negative right, see, for example, Sandberg (2003) or Boire (2001).

18

Pharmacological Enhancement of Cognition

Danielle C. Turner

Some believe that within 20 years people will have embraced the routine use of drugs to enhance performance. Smart drugs will become an essential addition to the armament of self-improvement techniques designed to give children the best possible start in life. 'Drug bars', where psychologists identify the most appropriate enhancer for individual clients, will burgeon on the high street, and most people will be consuming smart drugs in order to succeed. Indeed, society's increasing expectations will have made it impossible to achieve success without drugs, and those few 'misfits' who refuse to conform or are unable to afford the latest formulation will be gradually forced to the fringes of society.

We already know that the use of drugs to improve cognitive functioning is increasing and that this phenomenon is occurring despite the fact that we do not yet fully understand the physiological or psychological effects of currently available cognitive enhancers. In the US, 29 million prescriptions for analeptic agents were dispensed in 2004, the majority of these prescriptions being for patients under the age of 16 (Governale and Kaplan, 2005).

Of these prescriptions, 14 million were for various forms of the drug methylphenidate (more often known by its trade-name Ritalin™). Modafinil, a novel wakefulness-promoting drug, accounted for a further 2 million of the prescriptions, with amphetamines making up most of the remainder. Many of these prescriptions were for the treatment of attention deficit hyperactivity disorder (ADHD), a highly prevalent disorder of childhood characterized by impulsiveness and difficulty maintaining attention. However, there is increasing evidence that many of these drugs are being used by people who do not meet the diagnostic criteria for a mental illness but who are resorting to drugs to succeed in a highly competitive world. Most of the evidence for the use of smart drugs 'off-label' (that is, for an indication they are not licensed for) currently comes from the US, although there are indications that this may also be happening in the UK. Reports from the National Institute on Drug Abuse in the US in 2004 suggested that as many as 1 in 40 eighth-graders (13–14 year olds) were abusing methylphenidate, with this estimate doubling to 1 in 20 for 17–18 year olds (NIDA InfoFacts, 2005). By the time people reached university, this proportion was even higher, with just over 8 per cent of undergraduates (1 in 12) reporting the illegal use of prescription stimulants. Much of this consumption was of stolen drugs. Even several years ago, between January 1996 and December 1997, it was estimated that almost 700,000 doses of methylphenidate were stolen in the US (Kapner, 2003). It is also clear that most young people are taking these drugs in order to succeed. Against expectations, the most common motive given by students for their use of stimulants is their desire to increase their concentration and alertness, and not to get high (Teter et al, 2005). The move towards pharmaceutical enhancement amongst healthy people is not hypothetical fantasy, it is already happening.

MOTIVATIONS BEHIND THE DEVELOPMENT OF COGNITIVE ENHANCERS

Numerous neuropsychiatric disorders, such as ADHD, schizophrenia, frontal dementia and Parkinson's disease, are now considered to be characterized by persistent cognitive impairments. Yet the benefit of targeted cognitive enhancement in the treatment of psychiatric illness has been recognized only recently, prompting a wide search for appropriate treatments. One significant example of such an endeavour is the MATRICS (measurement and treatment research to improve cognition in schizophrenia) initiative of the National Institute of Mental Health (NIMH) in the US.

Effective cognitive functioning typically involves numerous neurotransmitter (brain chemical) systems and neuronal pathways, with several distinct neurotransmitters being implicated in the psychopharmacological enhancement of cognitive function (Robbins et al, 1997). 'Higher-order' executive functions, such as the ability to make decisions, plan, problem-solve and adapt behaviour, are crucial for the successful performance of many everyday procedures, like prioritizing tasks and remembering important information while doing another task (Stuss and Levine, 2002). Neurotransmitter systems can be manipulated with pharmacological agents to alter cognitive functioning. Several recent studies from our laboratory and others have shown that it is possible to improve cognitive functioning in a wide variety of groups of patient that suffer from cognitive impairments (Cardenas et al, 1994; McDowell et al, 1998; Mehta et al, 2000a; Aron et al, 2003; Turner et al, 2004a and 2004b; Rahman et al, 2005; Turner et al, 2005; Turner, 2006; Turner and Sahakian, 2006).

Understanding the mechanisms by which many of these drugs act in the brain to improve cognition is another major focus of scientific endeavour. This research aims to enable a greater understanding of the neural mechanisms of normal cognition. For example, neuroimaging in healthy volunteers has shown that methylphenidate, which has some cognitive enhancing properties (Elliott et al, 1997), can enhance processing efficiency within discrete neural networks in the brain, including the important frontal areas that are essential for higher cognitive functions (Mehta et al, 2000b). Through a greater understanding of the neurochemical mechanisms of memory and attention, it will become increasingly possible to selectively develop drugs that will help patients to lead effective and high-quality lives, unhindered by cognitive impairments.

Treatment of symptoms in a clinical population rarely poses serious ethical quandaries when there is clear benefit to the patient. What is more ethically challenging is the potential for cognitive improvement in healthy individuals with drugs developed to treat patient groups. Results showing improvements in healthy people arise from research into the effects of these drugs on healthy volunteers. Volunteer studies allow a comprehensive profile of a drug to be built, free from any confounding pathology that might be present in a patient population. While this work in healthy control groups is vital in furthering the pharmacological understanding of the mind and underlying brain mechanisms, it is also contentious, owing to the ethical issues surrounding performance enhancement of the healthy brain.

This has become more relevant in recent years with advances in drug design. Until recently, psychotropic medications had significant risks and side-effects that made them attractive only as an alternative to illness or disorder, when the benefits to the patient were considered to outweigh the side-effects. However, it is now becoming possible to pharmacologically enhance cognition with minimal side-effects in healthy volunteers (Elliott et al, 1997; Turner et al, 2003a). As part of a research programme to identify cognitive enhancers for patient use, our laboratory showed that a single dose of modafinil (a drug licensed for the treatment of narcolepsy) induced reliable improvements in short-term memory and planning abilities in healthy adult male volunteers (Turner et al, 2003a). Such newer drugs are typically developed to treat a medical condition and, although more work is needed to determine if these drugs will maintain their beneficial effects with chronic administration, many are increasingly being used for indications other than those they are licensed for (Farah et al, 2004). Despite this enthusiasm for the use of cognitive enhancers, however, their effects are not always reliably enhancing. It is also apparent with certain drugs, such as methylphenidate, that their enhancing effects may be restricted to a subpopulation of people with specific baseline performance levels.

THE VARIABLE EFFECTS OF COGNITIVE ENHANCERS

It is perhaps not surprising that cognitive enhancers do not act in the same way in everyone. For example, methylphenidate has distinct effects in different population groups and is associated both with enhancements and with impairments. Thus in elderly people (Turner et al, 2003b) the cognitive effects of the drug seen in younger people are grossly attenuated (Elliott et al, 1997; Rogers et al, 1999; Mehta et al, 2000b). Elliott et al (1997) found improvements in spatial span performance – in addition to spatial working memory – in young healthy volunteers after taking methylphenidate, although only if the participants received the drug in the first session. ADHD patients, on the other hand, despite being considerably impaired compared to normal volunteers (Turner, 2005), showed no improvement on spatial span after taking methylphenidate (Turner et al, 2005).

Some inconsistent results can be explained in terms of novelty. Improvements in planning seen with methylphenidate in healthy volunteers are typically seen when the task is novel and impairments seen when the task is familiar. It is also possible that inverted U-shaped functions, illustrating the Yerkes/Dodson principle of optimal levels of arousal for effective performance (first described by Yerkes and Dodson in 1908), are implicated in these different effects. For example, Hockey (1973) showed that exposing human volunteers to a loud noise improved reaction times on well-rehearsed or simple tasks, but impaired performance on more complex tasks. Evidence from animal studies indicates that the cognitive deficits observed during stress can result from excessive stimulation of dopamine D1 receptors in the brain, and that different tasks may require differing levels of prefrontal monoaminergic neurotransmitters for optimal performance (Arnsten and Robbins, 2002). Dopamine levels, and hence methylphenidate effects, might differentially affect the functioning of neural loops to engage preferentially in different forms of motor and cognitive functioning. In other words, the administration of methylphenidate, while optimizing performance on some tasks, would be detrimental to the performance of other tasks. This would account for the fact that healthy volunteers made more errors on one aspect of a test of cognitive flexibility (intradimensional errors) but made fewer errors on a conceptually different part of the test (extradimensional errors) following methylphenidate (Rogers et al, 1999).

One prediction of inverted U-shaped descriptions of catecholamine function is that baseline levels of performance, particularly on working memory tasks, may have some predictive value for the performance effects of drug administration. It is thus possible that differences in baseline span might also account for differences in response to methylphenidate. Mehta et al (2000b) showed that the beneficial effects of methylphenidate on working memory in normal adult volunteers were greatest in those volunteers with lower baseline working memory capacity. However, the opposite effect was observed in children with ADHD, where this time it was

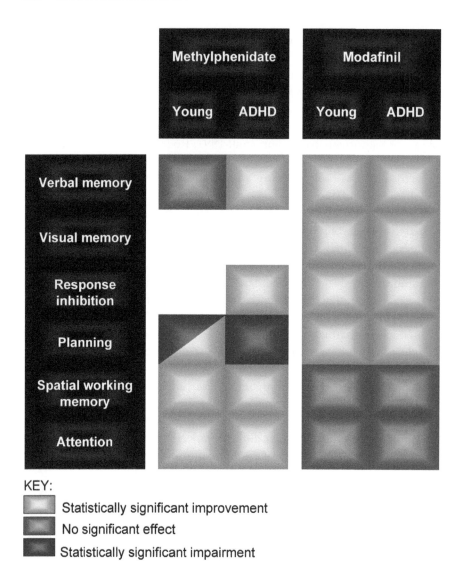

KEY:

Statistically significant improvement

No significant effect

Statistically significant impairment

Figure 18.1 *Effects on cognition of methylphenidate and modafinil*

Note: Two cognitive enhancers, methylphenidate and modafinil, produce very different effects on cognition compared to each other, both when given to young healthy volunteers ('young') and when given people with ADHD.

those with the highest baseline digit span scores that demonstrated the greatest improvement in spatial working memory following methylphenidate (Mehta et al, 2004). Baseline cognitive performance might thus provide clues as to which individuals might benefit the most from stimulant treatment.

In contrast to methylphenidate, modafinil seems to show a much more consistent effect across the different populations (although it should be noted here that fewer similar studies have been conducted)

In both young healthy volunteers and patients with ADHD, modafinil appears to enhance adaptive response inhibition (Turner et al, 2003a and 2004a). This effect of modafinil is similar to that found with methylphenidate (Tannock et al, 1989; Aron et al, 2003) and might account for the therapeutic efficacy (as indexed by clinical measures) of modafinil seen in ADHD (Boellner et al, 2006; Greenhill et al, 2005; Rugino and Copley, 2001; Swanson et al, 2006; Taylor and Russo, 2000). Modafinil also appears to enhance performance accuracy, accompanied by a slowing of response latency compared to placebo. This is particularly evident on tests of planning, where both healthy volunteers and patients with ADHD (and to a certain extent patients with schizophrenia) took longer to select their responses on the drug but were significantly more accurate, particularly at more difficult problems (Turner et al, 2003, 2004a and 2004b). Modafinil may therefore serve to improve accuracy by causing an increased tendency to evaluate a problem before initiating a response. This evaluative effect accords with current theories of a conceptually different form of impulsivity known as reflection impulsivity (Evenden, 1999), in which performance is impaired due to a deficit in utilizing all available information before making a decision.

In contrast to these common effects, modafinil also has some effects that are not consistent across groups. Thus no effect is noted on response inhibition in patients with schizophrenia, despite modafinil significantly improving the inhibition of prepotent responding in healthy adults and those with ADHD. Modafinil also significantly improves performance on a test of cognitive flexibility in schizophrenia, an effect not seen in the other groups. Similarly, sustained attention is improved in patients with ADHD (Turner et al, 2004a), an effect not seen in healthy volunteers (Turner et al, 2003a), suggesting that modafinil has the capacity to improve additional cognitive domains when baseline performance is impaired. A small study of modafinil in elderly volunteers (Randall et al, 2004) showed that it can also cause impairments. Elderly volunteers made more errors on a test of cognitive flexibility following 200mg modafinil (although not 100mg modafinil), in contrast to the improvements on this task seen in patients with schizophrenia. No effect of modafinil was seen in this elderly group on tests of visual memory, planning or sustained attention. It is possible that this absence of effects is due to similar changes noted in elderly patients receiving methylphenidate, where age-related changes in brain functioning could be accountable for the reduced effect of stimulant treatment.

It is apparent from the above overview that methylphenidate and modafinil have distinct cognitive enhancing effects, and therefore have different potentials for use in the different populations. Methylphenidate was effective in targeting cognition in adult ADHD (Turner et al, 2005) but caused some impairment in elderly volunteers (Turner et al, 2003b), young volunteers (Elliott et al, 1997), patients with schizophrenia (Carpenter et al, 1992; Levy et al, 1993; Szeszko et al, 1999) and children with ADHD (Mehta et al, 2004). In contrast, modafinil was effective in young volunteers (Turner et al, 2003a), adult ADHD patients (Turner et al, 2004a) and patients with schizophrenia (Turner et al, 2004b) but appeared to cause some impairments in elderly volunteers (Randall et al, 2004).

Cognitively, methylphenidate appears primarily to affect spatial working memory and planning as well as enhancing sustained attention, although these effects are dependent on novelty (Elliott et al, 1997) and baseline capacity (Mehta et al, 2000b and 2004). In contrast, modafinil has robust effects on verbal memory, visual memory and planning, but does not affect spatial working memory or sustained attention (it did improve sustained attention in ADHD patients, although not reliably). Modafinil also improves response inhibition in young volunteers and ADHD patients (although not in patients with schizophrenia) and appears to exert its effects by enhancing reflection. In contrast to methylphenidate, modafinil does not seem to have beneficial effects that depend on novelty or enhanced arousal, although it might possibly be exerting some effect on brain regions adaptive to practice. The two drugs do, however, appear to modulate certain distinct functions in similar ways, such as planning and response inhibition, providing evidence of some common underlying effects.

These similarities and differences are perhaps explained by the pharmacological mechanisms of action of methylphenidate and modafinil. Stahl (2002a and 2002b) has proposed an interesting hypothesis in which two distinct types of arousal, each mediated by separate pathways and transmitters, account for the differences seen between conventional stimulants and drugs such as modafinil. He has proposed that monoaminergic projections from the brainstem to the cortex are responsible for a type of arousal that involves tense hyperarousal and stimulated vigilance. This type of arousal is mediated by the monoamines dopamine, noradrenaline, serotonin and acetylcholine, with stimulants such as amphetamine, methylphenidate and caffeine activating this system. In contrast, the second form of arousal is postulated to be a more reflective type of calm wakefulness, associated with the natural sleep–wake cycle, in which there is internal vigilance to executive functions with focusing on cognitive tasks. Stimulants such as amphetamine and caffeine are thought to exert their effects on both the catecholaminergic and histaminergic systems, while modafinil might instead be acting selectively through the second pathway (Stahl, 2002a and 2002b).

LOOKING TO THE FUTURE

Research examining cognitive enhancement generally benefits from the assessment of the pharmacological effects of drugs on healthy volunteers, where the results are not confounded by pathology. This often raises difficult ethical considerations, however, not least of which is that, as society searches for mechanisms to improve performance and memory, safety (and the potential for 'compensatory' impairments, as described above) becomes of particular concern if a pharmacological agent is to be used to enhance, rather than to treat, performance. Patients with severely debilitating symptoms will often tolerate the side-effects of drug treatment, because improvements in symptoms outweigh the negative aspects of treatment. Furthermore, it is very difficult to be certain about the potential for subtle, rare or long-term side-effects, and thus a full exploration of the long-term implications of any treatment that might be used by the healthy population is imperative.

There are also several other 'neuroethical' issues related to the impact on society of widespread cognitive enhancement, many of which have been raised in this volume and to which there are currently very few answers. These include hypothetical situations where people might be pressurized into enhancing their cognitive abilities (for example by employers or teachers), issues relating to unfair distribution of cognitive enhancing agents, and philosophical issues relating to the meaning of self (and what it means to be a person and to value human life in all its imperfections). Some of these concerns arise directly from scientific findings such as the demonstration that methylphenidate is likely to benefit only those people whose baseline performance falls within a narrow range.

Ultimately, however, it is likely to be the wider population that decides whether these drugs should be embraced or banned. Society needs to decide whether, for example, people should be striving to make their brains function better than normal by using drugs. One mechanism to achieve this is through the provision of reliable information. Indeed, it has even been suggested that universities should start training 'neuroeducators' – people able to advise parents and schools on how to incorporate developments in neuroscience, such as cognitive enhancement, into education. A number of UK universities, including Cambridge, are already offering courses that combine neuroscience and education. Hopefully, through continued discussions relating to new neuroscientific technologies, we will move forward in a considered way that maximizes the benefit these developments can bring while minimizing any harms.

19

The Economics of Brain Emulations

Robin Hanson

INTRODUCTION

Technologists and economists both think about the future sometimes, but they each have blind spots. Technologists think about specific future technologies, which they may foresee in some detail. Unfortunately, technologists then mostly use amateur intuitions about the social world to predict the broader social implications of these technologies. This makes it hard for technologists to identify the technologies which will have the largest social impact. Economists, in contrast, have a professional understanding of the social world, and are well positioned to analyse the social implications of specific technologies. Using simple mathematical models based on powerful general concepts, economists could go well beyond simple trend projections. Unfortunately, economists mostly rely on amateur intuitions about the feasibility of future technologies. Substantial technical innovations often seem to them like science fiction and too silly to take seriously. Economists' future projections thus usually ignore specific future technologies.

As an economist (a tenured professor) with a technology background (a physics master's degree and nine years of computer research), I try to avoid these blind spots. By applying economic theory to specific future technologies, I hope to go beyond trend projection to foresee the social consequences and relative importance of future technologies.

Of the many future technologies I have considered over the years, one stands out to me as likely to have the largest impact: brain emulations. This technology also happens to be relatively easy to analyse with standard economic tools. But before discussing this technology, let me outline an independent reason we have to expect a huge economic transition in the next century.

LONG-TERM TRENDS

A postcard summary of life, the universe and everything might go as follows. Our universe appeared and started expanding. Life appeared somewhere and then on Earth it began to make larger and smarter animals. Eventually human beings appeared and became smarter and more numerous by inventing language, then farming, then industry and most recently computers.

The events in this summary are not evenly distributed in time. The first events are relatively evenly distributed: the universe started 14 billion years ago, life appeared by 4 billion years ago, and on Earth animals started growing larger and smarter about half a billion years ago. But by comparison the other events are very closely spaced: our species appeared a few million years ago, farming started about 10,000 years ago, industry started about 200 years ago and computers started a few decades ago.

One might worry that this list is a biased sample of globally important events, because we human beings overemphasize events that are about us. But I think otherwise. I think these are in fact the globally important events, because they separate a chain of distinct key exponential growth modes. Exponential growth is where some quantity doubles after a certain time duration, and then continues to double again and again after similar durations. At each point in history some crucial quantity has been growing exponentially. And at a few rare transition points, the growth rate has suddenly increased.

The slowest growth mode started first. Our 14-billion-year-old universe is expanding, and that expansion is now roughly exponential due to a mysterious 'dark energy'. The distance between the galaxies is predicted to double every 10 billion years.

We don't know enough about the history of non-animal life in the universe to identify its growth rates, but we can see that for the last half billion years the size of animals on Earth has been growing exponentially. While the size of the typical animal has changed little, the variation among animal sizes has greatly increased. Because of this, the mass of the largest animal has doubled about every 70 million years, and the mass of the largest brain has doubled about every third of a 100 million years. So the largest brains have doubled about 300 times faster than the distance between galaxies.

Human beings (really our 'human-like ancestors') began with some of the largest brains around (relative to their bodies), and then tripled their brain size. Those brains, and the innovations they embodied, seem to have enabled a huge growth in the human niche – the planet supported about 10,000 humans about 2 million years ago, but about 4 million humans about 10,000 years ago.

While data is scarce, this growth seems roughly exponential, doubling about every 200,000 years. This is 150 times faster than animal brains grew. (This growth rate for the human niche is consistent with faster growth for our ancestors, as some groups killed off others to take over the niche.)

About 10,000 years ago, those 4 million human beings began to settle and farm, instead of migrating to hunt and gather. The human population on Earth then began to double about every 900 years. This farming humans growth rate is about 250 times faster than that of the previous hunting humans.

Since the industrial revolution began a few hundred years ago, the human population has grown even faster. Before the industrial revolution, total human wealth grew so slowly that population quickly caught up, keeping wealth per person near subsistence level. But in the last century or so wealth has grown faster than population, allowing for great increases in wealth per person.

Economists' best estimates of total world product (average income per person times the number of people) show it has been growing exponentially over the last century, doubling about every 15 years, or about 60 times faster than under farming. And a model of the whole time series as a transition from a farming exponential mode to an industry exponential mode suggests that the transition is not over yet – we are slowly approaching an income doubling time of about six years, or 150 times the farming growth rate.

A revised postcard summary of life, the universe and everything, therefore, is that an exponentially growing universe gave life to a sequence of faster and faster exponential growth modes. First the largest animal brains grew slowly, and then the wealth of human hunters grew faster. Next farmer wealth grew much faster, and finally industry wealth grew faster still. Perhaps each new growth mode could not start until the previous mode had reached a certain enabling scale. That is, perhaps human beings could not grow via culture until animal brains were large enough, farming was not feasible until hunter populations were dense enough and industry was not possible until there were enough farmers near each other.

Notice how many important events are left out of this postcard summary. Fire, writing, cities, sailing, printing presses, steam engines, electricity, assembly lines, radio and hundreds of other key innovations are not listed separately here. The reason is that most big changes are a *part* of some growth mode, but do not cause an *increase* in the mode's growth rate. While we do not know what exactly has made growth rates change, we do see that the number of such causes observed so far can be counted on the fingers of one hand.

While growth rates have varied widely, growth rate changes have been surprisingly consistent – each mode grew from 150 to 300 times faster than its predecessor. Also, the recent modes have made a similar number of doublings before giving rise to a new mode. While the universe has barely completed one dark-energy doubling time, and the largest animals grew through 16 doublings, hunting grew through nine doublings, farming grew through seven and a half doublings and industry has so far completed a bit over nine doublings.

This pattern explains event clustering – transitions between faster growth modes that double a similar number of times *must* cluster closer and closer in time. But looking at this pattern, we should wonder: are we in the last growth mode, or will there be more?

A NEW GROWTH MODE?

If a new growth transition were to be similar to the last few, in terms of its number of prior doublings and its increase in the growth rate, then the remarkable consistency in the previous transitions allows a remarkably precise prediction. A new growth mode should arise sometime within about the next seven industry mode doublings (in other words about the next 70 years) and give a new wealth doubling time of between roughly one and two weeks.

How sudden would such a transition be? We only have transition data on the last two transitions, and of those the industry transition was smoother. If the next transition happened around 2040, and was as smooth as the industry transition, then a simple model predicts the sequence of expected annual percentage growth rates to be 6.1, 6.1, 6.6, 8.0, 14, 41, 147, 475, 1025 and so on. If growth rates fluctuate by about 0.5 per cent per year, then growth rates would have doubled within two years of the first noticeable change, and within two more years the world economy would be doubling more frequently than yearly.

The suggestion that the world economy will soon double every week or two, after a transition lasting only a few years, seems so far from ordinary experience as to be, well, 'crazy'. Of course, similar predictions made before the previous transitions would have seemed similarly crazy; nevertheless, it seems hard to take this scenario seriously without at least some account of how it could be possible.

The first point to note is that we should not expect to be able to get a very detailed account of a new growth mode. After all, most economics has been designed to explain the actual social worlds that we have seen so far, and not all the possible social worlds that might exist. And we are still pretty ignorant about the fundamental drivers of the previous modes. But we do want at least a sketchy account. Of the many future technologies that technologists have forecast, which could plausibly have anywhere near this impact on the economy?

One helpful hint is that innovations in larger economic sectors can produce larger social impacts. In the US we spend about 1.5 per cent of income on farming, 1.5 per cent on mining, 2 per cent on gas and electricity, 2.5 per cent on communications, 3 per cent on transportation, and 3.5 per cent on construction. These small fractions make it hard to see how even dramatic innovations in these sectors could induce much faster growth. For such drama, we must look beyond the usual technology favourites, such as space colonization, fusion energy, air cars, sea cities or picture phones. We probably must even look beyond radical nanotechnology – while this might dramatically reduce the cost of capital for manufacturing, we only spend about 5 per cent of income there.

A more promising fraction is the 70 per cent of income we now pay for human labour, as opposed to other kinds of physical and social capital. Greatly lowering this cost could have a huge impact. And a robotics or artificial intelligence technology good enough to substitute wholesale for most human labour might just greatly lower such costs.

Brain Emulations

For centuries now, people have been concerned about the possibility of machines replacing human labour. And many kinds of labour have in fact been replaced by machines. At first machines replaced people at tasks needing physical strength, but more recently machines have replaced them at mental tasks.

On the whole, however, machines have mainly helped people be more productive at tasks that machines cannot do. By complementing human beings, machines have so far greatly raised the value of most human labour. Because of this, most economists have not worried about machines replacing people.

Previous trends need not continue, however. The key point to understand is that while *tasks* complement each other, *individuals* are substitutes for doing each task. There are many tasks that we want done, and machines are better suited to some tasks than to others. Thus slowly improving machines have two effects on human labour. First, machines get better at the tasks machines do best, which makes doing all the other tasks well more valuable. This complementary effect raises the demand for human labour. And second, some marginal tasks switch from people to machines. This substitution effect lowers the demand for human labour.

So far, people still do most tasks worth doing, and so the net effect has been to raise human wages. But this picture would change dramatically if we had machines that were good at almost all the tasks people now do. Human wages could then fall with the falling price of machines. And since the number of machines could grow as fast as the economy needed them, human population growth would no longer limit economic growth. In this scenario, simple growth models can easily allow a new doubling time of a month, a week or even less.

Now admittedly, progress in robotics and artificial intelligence has been slow; it has been hard to write capable software. At current rates of progress it could be centuries before machines could get good at almost all tasks that people do. There is one approach to broadly capable artificial intelligence, however, that seems likely to succeed within the next century: brain emulations.

The idea here is to not 'write' the relevant software, but to 'port' it from a real human brain. Take a brain, and scan it in enough detail to see each neuron's type and its connections to other neurons. Study each type of neuron in enough detail to create a computer model of how its output signals depend on its input signals. Finally, create a computer model of the entire brain, connecting together models of each neuron, and connecting them to emulated eyes, ears, mouth and so on.

If the connection information and the neuron models are good enough, then the brain model should have roughly the same input–output behaviour as the original brain. That is, you could talk to it and it would talk back. And if you could convince it to work for you, it could accomplish most of the same sorts of tasks as the original brain. It might even be conscious and enjoy its life (though this claim may long remain disputed). And once you had one such brain, you could make billions more by just copying the software.

Three technologies are needed to make this work: enough neuron-type models, fast enough scanning and large enough computers. These technologies have been steadily improving for many decades now, and they each seem likely to be ready by mid-century, if not by quarter-century. Of the three, progress in neuron modelling seems the hardest to forecast. But we already have good enough models of many neuron types, we already have slow but accurate enough scanners, and computers should be fast enough in a few decades. No grand breakthroughs seem required, just hard work and steady progress of the sort we have already seen.

Thus brain emulations seem likely to appear in time to cause the next big growth mode, and simple economic models suggest they are capable of producing such a mode. Within a few years human wages could begin falling dramatically, while economic growth rates skyrocket.

More precisely, such changes could happen if they were allowed. To keep our models simple and comparable, economists usually start by modelling peaceful low-regulation scenarios, such as where wages are set by supply and demand, and where people could make as many brain emulations as they wanted. One can imagine regulatory action in the form of wealth transfers to ensure that human beings do not starve due to falling wages, and minimum wages or population controls to limit the number of brain simulations. Given the lack of a strong world government, however, it is not clear whether such regulations would be feasible, or if feasible whether they would be desirable. As important as these questions are, they will have to wait for another essay.

Part VII

Happier?

20

Happier:
A Psychopharmacology Perspective

David Nutt

Happiness can be viewed from a number of neuroscience perspectives, including those relating to the actions of various neurotransmitters and hormones. It also is a factor in the use of mind-altering or addictive drugs and an element in a number of psychiatric disorders.

In some circles it seems to be believed that we have already achieved the goal of 'happy pills' – a pejorative term that some parts of the media use for antidepressant drugs (see Nutt, 2003). In reality antidepressant drugs do not make normal people happy – they simply, and over a period of weeks, lift the depressed mood in those whose genes and/or life stresses predispose to this disabling condition.

A PSYCHIATRIC PERSPECTIVE

Psychiatry can give us an interesting take on the question of happiness. The spectrum of mood than runs from depression through normality to mania can be viewed as a happiness dimension. In depression it is difficult to experience happiness, either in response to positive life events or even in remembering happy episodes in the past. Conversely, in some people with mania there is a pervasive increase in seeming happiness that is often associated with an unfounded sense of self-confidence and optimism. In extreme cases, grandiosity that spills over into quasi-religious beliefs of divine calling with special powers can be manifest.

We can therefore view this bipolarity of mood as giving some insights into happiness. In depression there is a pathological inability to experience happiness, as well as other changes in mental function such as altered sleep, appetite, libido and energy. In mania the opposite situation prevails – there is a pathological level of happiness or reactivity to events in a pleasurable way that can lead to highly inappropriate and dangerous behaviour. But there is one other important aspect to the nature of happiness in both depression and mania – the lack of insight into the essentially abnormal nature of the mental state. This may tell us something significant about the primary driver of the change in happiness in each condition, though there is little research into this issue.

NEUROTRANSMITTERS AND HAPPINESS

If altered mood relates to altered experiences of happiness, can an understanding of the neuroscience of mood mechanisms lead to insights into the nature of happiness and, probably more importantly, ways to improve the amount of it in society?

Both depression and mania have been the subject of intense neuroscience research for over 50 years, and there are some facts to guide us. Starting with depression, we know that it may be caused by stress and certain drugs and is mimicked by grief. We also know that antidepressants and other treatments, for example ECT (shock therapy), alleviate depression effectively. This knowledge offers two approaches to happiness: the reduction of stress and the elevation of mood with drugs or other proven antidepressant treatments.

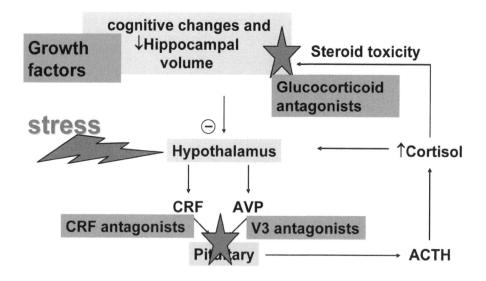

Figure 20.1 *Hormone stress axis: Targets for new treatments*

PREVENTING STRESS

The mechanisms of stress are well researched (see Figure 20.1). In brief, psychological stressors (threats) as well as physical stressors (pain, injury, blood loss and so on) act to stimulate the release of two peptide neurotransmitters in the hypothalamus, CRF and AVP, that together stimulate the release of the hormone ACTH from the pituitary gland. The released ACTH then travels in the blood to the adrenal glands, where it stimulates the release of cortisol, the stress-protective hormone. As cortisol levels in the blood rise, it begins to stimulate receptors in the brain and pituitary that switch off the release of CRF and AVP – this is called negative feedback and it serves to keep the system in homeostatic control.

In some circumstances, however, either after prolonged excessive uncontrollable stress, or in people with early life stressors or genetic predisposition, the feedback control of this system is disrupted, leading to excessive cortisol production. This in turn can damage certain parts of the brain, such as the hippocampus, leading to more dysregulation of the system and effects on other brain processes, especially depression and memory problems. Understanding the component parts of the stress process means that we can develop interventions that might work to reset it once it goes wrong – in other words to lift depression – and perhaps even to buffer it against stress in the first place – a form of prophylaxis. Such potential new drugs include antagonists (blockers) of CRF and AVP receptors as well as glucocorticoid receptor antagonists (cortisol receptor blockers) that might prevent cortisol damaging the brain. Such drugs are in testing stages at present and are likely to be available for study and treatment within the next ten years, assuming they are shown to be effective (Holsboer, 1999).

MIMICKING ANTIDEPRESSANTS

The mechanisms of action of most antidepressants and ECT is now partially understood, with increases in brain serotonin or noradrenaline function being the two main alternative pathways to restore low mood to normal (Slattery et al, 2004). In the light of these observations, it has been suggested that treatments that increase or stabilize especially serotonin (5HT) levels in the brain might have the ability to protect against depression and other stress-related disorders. Precursors of 5HT, especially 5HTP (5-hydroxy tryptophan), are sold in heath food shops and there is some evidence that they may be of value in preventing depression. Conversely, certain behaviours tend to lower 5HT in the brain, and these include dieting, especially low-protein diets (as the amino acid that 5HT is synthesized from, L-tryptophan, is only found in protein).

Based on this knowledge, it may be possible to improve the diet of individuals to optimize 5HT synthesis in the brain and so reduce their risk of depression, hence enhancing their periods of happiness. For example, fortifying milk and other dairy products with L-tryptophan has been explored and may yet prove a useful public health measure.

Other approaches to happiness may emerge from the developing disciplines of extra- and intra-cranial brain stimulation. Extra-cranial brain stimulation uses transient magnetic pulses in the form of transcranial magnetic stimulation (TMS) to produce localized activation of brain regions just under the skull. It has been shown to lift mood in depression, perhaps by releasing dopamine (Keck et al, 2002), and some have speculated that in theory it might be used to induce happiness or at least a sense of wellbeing in normal people. Intra-cranial stimulation uses implanted electrical wires that can be turned on from outside the brain to locally stimulate brain regions or nerve tracks. It has been shown to be a valuable treatment in certain neurological disorders, being especially useful in stopping the tremor in Parkinson's disease and the muscle spasms in some forms of dystonia. As the surgical procedures have become better established, this approach has been applied to treat depression and obsessive compulsive disorder (Mayberg et al, 2005). In some of these cases, as in the patients with Parkinson's disease, an elevation in mood has been reported; indeed occasionally mood has even become inappropriately raised, leading to unwanted behaviours, especially gambling. This effect is reminiscent of the classic animal studies by Olds and Olds in the 1950s, which found that rats would administer electric currents into certain brain regions in preference to all other behaviours. This so-called self-stimulation circuit turns out to be a dopamine pathway that is implicated in drug addiction and mood disorders.

DOPAMINE AND HAPPINESS

The neurotransmitter with the best evidence for an involvement in mood and happiness is dopamine. Drugs that cause the release of dopamine, especially the stimulants (cocaine, amphetamines and nicotine) can give a sense a pleasure – a high – that is very rewarding to the user and which perpetuates repeated drug use. They also increase energy and drive, and can push susceptible individuals into a state of excess energy and drive such as mania. Figure 20.2 suggests there may be a staircase of the effects of dopamine, from normal contentment to the pathological pleasure of mania. Conversely, drugs that block dopamine function such as the major tranquillizers, for example haloperidol, are associated with the induction of depression in patients and also in transient lowering of mood in normal volunteers. Similarly, reducing the production of dopamine by limiting the amount of the precursor that gets into the brain can reduce mood, block the effects of amphetamine and has been used as a treatment for mania (see McTavish et al, 1999 and 2001).

The drawback of dopamine as a target for happiness-promoting drugs is that the response may not always be regulated – in some people excess actions could lead to mania, with its attendant dangers. Also, because dopamine is so intimately

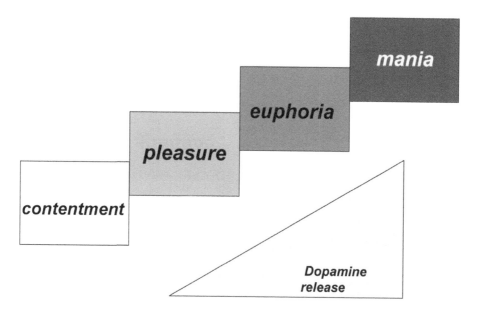

Figure 20.2 *Stairway to heaven?*

implicated in movement as well as mood (loss of certain dopamine cells in the brain leads to Parkinson's disease), increasing its action can lead to unwanted stereotyped movements such as tics. In some cases, dopamine enhancement can also lead to mental stereotypes, such as compulsive repetitive thoughts or ideas, which can be upsetting.

Moreover, there is the important question as to whether increasing dopamine function in the brain leads to a 'true' happiness, or is rather one dimension (that of pleasure and drive) in a complex set of vectors that together might underpin happiness.

OTHER HAPPINESS NEUROTRANSMITTERS

Drugs that act on neurotransmitter systems other than dopamine can also lead to states of altered happiness. Some would say that this form of happiness is less to do with reward and energy and more to do with inner tranquillity – perhaps best summed up as transcendence (see Figure 20.3). Opiates such as morphine and heroin are well known to do this, as they powerfully mimic the brain's own opiate system – the endorphins. However, because they do this so strongly, their pleasurable effects are seen only in the short term, as problems of dependence and withdrawal develop on repeated use. Cannabis may do the same, but again

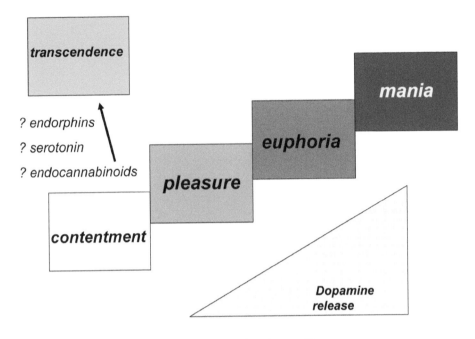

Figure 20.3 *Another path?*

problems of dependence may ensue. LSD and other psychedelics can produce in some people a sense of intense personal understanding and contentment that could be called blissful happiness. These are probably acting via the 5HT system. MDMA (ecstasy) is another drug that can make people happier and more at ease with others in social settings. Ecstasy works by releasing 5HT and probably also dopamine. (For more details on these drugs, see Lingford-Hughes and Nutt, 2003, and chapters in Nutt et al, 2006.)

Finally, we must not overlook the near-universal mood-altering drug: alcohol (ethanol), which induces happiness by enhancing brain GABA function (GABA is the main inhibitory – calming – neurotransmitter in the brain). We think that alcohol makes people happy partly by reducing anxiety, the reason it is used to lubricate social interactions such as parties and deal with fear of flying, and partly because it directly releases endorphins and perhaps dopamine and other neurotransmitters.

THE FUTURE?

Given the considerable knowledge we now have about the brain mechanisms of mood, stress and to a lesser extent happiness, are we on the path to the discovery

Figure 20.4 *Regulatory changes for happiness drugs: Could they ever be marketed?*

of new, better and safer happiness-promoting drugs? In theory the answer is yes. The pharma industry continues to make regular progress in drug treatments based on neurotransmitter science, through a well-defined process of synthetic chemistry, with pharmacological screening against targets of interest. However, in the case of happiness drugs, the situation is complicated by external regulatory factors that militate against the investment, most importantly whether they could ever sell such drugs.

The basis of this confusion in shown in Figure 20.4, where it can be seen that happiness drugs would fall between two very different regulatory systems – one for legal medicinal compounds, the Medicines Act, and one for illicit or illegal drugs, the Misuse of Drugs Act. A happiness drug could currently not be licensed and sold as a medicine, because it would not be treating a medical disorder – so companies would not be able to sell it and recoup their investment. Their only recourse would be to try to license it as an antidepressant, but happiness enhancers might not be such good antidepressants as the current crop, so might not be licensed. Moreover, companies could not market the drug outside its licensed indication.

If the drug could only be made available outside the Medicines Act, then it would have to be sold illegitimately or as a foodstuff/commodity. The first option is probably a non-starter, as no company would make a drug that could only be sold on the 'street', though the growth of internet sales of drugs might mean that small firms might take the risk, as some individuals do now in selling dance and other drugs online.

In principle a happiness drug could be regulated as a commodity, like alcohol or tobacco, but to do this would require a significant change in the attitude of parliament, though the recent move to ban public smoking may suggest that such approaches would not necessarily be out of the question.

The current regulatory situation is therefore both arbitrary and restrictive of new developments, problems discussed in the Foresight report 'Brain science, addiction and drugs' (Foresight, 2005) and addressed in more detail in the recently published Academy of Medical Sciences report (2008). This gives a good critique of the philosophical, practical and public issues relating to recreational drug use and opens the possibility of a more open and balanced approach to new drugs that may improve wellbeing, including happiness.

21

What's Your Mission in Life? Why Being Happy Should *Not* Be Your Priority

Nick Baylis

INTRODUCTION: BEING HAPPY SHOULD NOT BE OUR PRIORITY IN LIFE

For a long time now, 'how to increase your happiness' has been a promise that's been packaged and sold as a consumer product, much like ice-cream or alcohol. If orange is the new black, then happy is the new rich, the new beautiful, the new 'sadness-free diet' promising a swift end to all our ills.

'I just want to be happy!' is the intoxicating mantra hanging thickly in the air of our everyday world. But this highly popular pursuit has seriously damaging consequences for us, because aiming to be happier is *not* what Nature intended as our priority in life. It's not what we're innately driven to make our central mission.

Then what on Earth is? The answer, I believe, is *progress*, or, more specifically, progress in our relationship with life. To this end, we all of us have a profound calling to forge all manner of healthy relationships, because the best way for us to survive and thrive is through teaming up and working in partnership. After all, a healthy relationship is one in which the partners achieve far more together than they could ever do alone. What's more, our inborn drive is not only for the healthier relationships we can have with the people around us, with our communities and the natural environment; it is also a drive to improve the harmony between our mind and body, our conscious and subconscious, and the relationship we have with ourselves. (For instance, do we encourage and care for ourselves, or do we spite ourselves and self-sabotage?) In essence, then, our inherited hunger for life is about making progress in our essential and fundamental relationships with ourselves and the world around us.

To aid us on our journey, our evolutionary history has also hard-wired us to be hungry for improvement in all the personal resources that will nourish our healthy relationships. Resources here include helpful strategies and skills, as well as beauty, brains, health and wealth. By developing our stockpile of such personal assets, our key relationships can thrive, which in turn make us better able to compete for our particular niche in life. That is what I mean here by 'progress'.

This is not a simple matter of more means better. Progress in relationships comes about by developing the skill and wisdom with which we gather, manage and deploy our personal assets. In other words, improvement in our life quality requires us to find ever more healthy ways to earn our resources, to nurture them and to invest them well in enhancing our relationships. By contrast, the ill-judged acquisition of resources (for example gaining extra income through self-damaging overwork), or the ill-judged expenditure of our resources (for example trying to buy friendship or gulp down pleasure), are the sorts of misguided approaches that mistake the means for the ends, and will only lead to a deterioration in life quality.

In the psychologically healthy individual, the inborn calling for progress is overwhelmingly strong. From wherever we are now in life, no matter how beautiful or clever or accomplished, no matter how downtrodden or injured or imprisoned, we still want to feel ourselves moving forward in some way. Our vocation for progress is deeply engrained in the make-up of our genes. Yes, our individual creativity and self-expression can determine which relationships we prioritize and how we go about doing so, but there's no escaping the fact that, to live a healthy life, prioritize them we must.

So, when pondering how future technologies can help to enhance and prolong human life, it's crucial we get our goalposts in the right place. Happiness is a mere corner flag. Healthier relationships is what we're aiming for.

What is the evidence for this?

A personal anecdote may serve as an introduction here. By an unfortunate oversight at St Albans Boys School in 1981, some classical subjects such as happiness, women, earning money and life-in-general were left off the curriculum. Either that or my comrades and I were bunking-off class when the careers teacher finally got around to them one Friday afternoon. In the years that followed, those gaps in my understanding of the art and craft of living life caused me quite a few problems. By the time I was 25, I was a hopelessly incompetent waiter wreaking havoc with bowls of soup and the cake trolley in unsuspecting restaurants.

It's these glaring shortfalls in my own understanding of life that led me falteringly, via Feltham Young Offenders Prison, to become part of the study of wellbeing – which is the systematic study of life going well. I therefore feel it's my job to explore and encourage such subjects as psychological, social and physical good health, as well as expertise and high performance, beauty in all its forms, thriving in the face of adversity, and living in harmony with the natural environment. This emerging field launched itself with a three-day conference hosted by the Royal Society in 2003 (Huppert et al, 2005), and is now represented by the newly formed Well-Being Institute of the University of Cambridge. Our movement seeks to incorporate beneath its flag the emerging science of positive psychology and all other disciplines (for example technology, art, medicine, architecture, sociology and economics) that can shed light on how we can help life to thrive and flourish for individuals and families, communities and nations. Our approach needs to be profoundly integrated and holistic if we're to see the bigger picture, because it's no accident that the word 'health' means 'wholeness' and the verb 'to heal' means 'to make whole'. Fortunately, the wellbeing portfolio already incorporates some excellent studies which have tracked hundreds of individuals across a lifetime or conducted field experiments and surveys (see, for example, Putnam, 2000; Vaillant, 2002; Morris, 2003; Baylis, 2005). From this accumulated evidence, it appears that the common characteristics of healthy, helpful and good-hearted lives are very notably those strategies that progress our fundamental relationships with ourselves and the world around us.

A definition of happiness and pleasure

By 'happiness' I mean that whole range of sensations, feelings and thoughts which bring us pleasure, whether that pleasure be fleeting or lasting, specific or general. While acknowledging that others may prefer alternative definitions, the way I'm using the word here, happiness is the catch-all for such feelings as joy or satisfaction, love or hope, confidence or pride. Indeed, from here on in I'm going to use the words happiness and pleasure interchangeably to mean one and the same thing: our feeling (or thought) that something (or everything) feels good.

What is the role of happiness if it's not our purpose in life?

The true role of happiness is to do one or all of three things:

1   Our pleasurable emotions can, if well used, be a *fuel* (a self-motivation) for improving our resources, for example if we feel confident enough, we will take on a challenge.

2   Pleasure can also be a *reward* for our improvements, for example we will feel satisfaction at having learned something valuable.

3   Pleasure can be a *flashing beacon* saying, 'Over here if you want to make progress!', for example a beautiful environment calls to our hearts to live there because it promises us health and prosperity.

But the crucial distinction is that our pleasurable thoughts and feelings are not in themselves the improvement. Pleasure can be the fuel, the reward and the guiding compass, but pleasure is not itself the progress, not the improvement in our resources. What we actually do – for example the activity, the challenge or the learning – is the progress.

The bottom line is that our human brain is only interested in our happiness because it is *one* means by which we get ourselves to make progress in our essential relationships with the world around us. Happiness is simply a means to that end; it's not an end in itself.

What happens if we mistakenly put 'maximizing pleasure' at the centre of our lives?

If we mistake pleasure for being the true end-goal of our efforts, we try to go straight to pleasurable feelings as directly as possible via, for example, alcohol and other drugs, over-eating, over-spending, internet pornography, too much TV, or escapist fantasy. We become 'passive voyeurs' of real life, couch potatoes gulping down pleasure in our attempts to feel good. But trying to increase our pleasure in ways that are unrelated to our real-life progress leads only to stagnation and deterioration. That's how we end up with addictions, excessive debt, obesity, eating disorders and all the other serious symptoms of this unhelpful prioritization of pleasure which characterize so many 21st-century cultures.

Even well-earned pleasure can derail us if we don't channel it helpfully

Even deeply pleasurable feelings that have been well earned through real progress can, if we don't channel them constructively, lead us into trouble. We can, for instance, in one way or another behave like the delighted young person who, not knowing quite what to do with their overwhelming sense of joy and relief at passing their exams, not only gets drunk but feels so attracted by their friend across the table, that they quite happily have unsafe sex. The following day, there's hell to pay for these ill-considered actions, and all this misery has come about because the individual couldn't quite cope with the overwhelming pleasure of their joy and relief at passing their exams. We probably all have our own anecdotes of how the sheer happiness of such feelings of love, hope, curiosity, pride or satisfaction caused our apple-cart to turn over because we didn't exercise sufficient skill to helpfully channel their powerful energy.

What happens if we try to minimize our pain?

Pleasure is only one half of our motivation for progress; pain *is* the other, not only physical pain, but our emotional anger, frustration, fear, shame, loneliness and sadness. The trouble is, if we mistakenly regard pleasurable feelings as being the highest purpose of our life (in other words the indicator that all is well), then we are prone to regard pain as our arch enemy and so avoid or quell our painful physical feelings or emotions by whatever means come to hand. As the following section makes clear, this is a mistake.

What is the role of pain, if it's not always negative?

Pain, if overwhelming or not channelled constructively, can hinder our progress in life by causing us to stop trying, to turn back or to become embittered. On the other hand, our painful emotions are an equally powerful source of energy and guidance to our pleasurable ones, and if we can learn how to harness their power and channel it in constructive directions, that emotional energy can help us progress in improving all the personal resources which help our key relationships to thrive and flourish.

More specifically, as the equal and opposite of pleasure, it is the role of pain to do one or all of three things:

1 Our painful emotions can, if well used, be a *fuel* (a self-motivation) for improving our resources, for example if we feel angry enough we will take on a challenge.
2 Pain can be a *punishment* (in other words a negative reward) for not making sufficient progress, for example we feel frustration at not having learned anything valuable.
3 Pain can be a *flashing beacon* saying, 'Over here if you want to make progress!', for example facing and then overcoming our dire fear of public speaking will allow us a sense of personal growth.

Pain should certainly not be regarded as a red stop-light or a skull and crossbones warning us to turn back. Indeed, the things that hurt and scare us most can often be the richest source of learning and progress if only we can confront them and find ways to channel them. For instance, consider those occasions when, out of shame or anger or loneliness, we have acted in very positive ways in an attempt to dramatically improve the situation and have succeeded in doing so. On those occasions that we have managed to turn our troubles into triumphs, we are not unlike Monty Roberts (affectionately known as 'the horse whisperer'), who took his childhood experience of the terrifying bullying at the hands of his brutal father and used this pain to help himself develop a gentle and respectful method to train wild horses. Alas, the alchemy that allows us to channel such pains to fuel something truly positive is a crucial skill in short supply.

Pleasure can lead to stagnation; pain can lead to progress – it all depends on how well we've learned to channel their energy

By studying whole lifetimes from birth to death, we can better see the workings of some of the most undervalued and misunderstood relationships: we can see that pains in childhood (such as setbacks and rejections), if well channelled, can lead to remarkable progress. And how pleasures (such as success and privilege), if poorly channelled, can derail life and lead to misery. It's not what happens to us that is the making or the measure of us, it's what we do about it. Only our skill at channelling determines the eventual outcome. It's as if we are all of us steam engines fuelled by a furnace that burns both coal (pain) and wood (pleasure) so as to create the energy of self-motivation. This means that pain is not our enemy. Feeling nothing at all is most often our enemy, because without energy, without drive, without self-motivation, there can be no progress.

So how exactly do we 'channel' our pains and our pleasures to help us make real-life progress?

In a nutshell, to constructively channel our pain, we need to *generate a sufficiently self-motivating journey and goal that in all likelihood will compensate us for the pains we have suffered.* In other words, we need to envisage a future for ourselves where things are so improved that we would actually be grateful for days things had gone wrong, because of the goodness that had eventually come from them. Once we can envisage such an attractive future, we then work backwards from it to where we are at the moment, so we can see what variety of routes might take us there.

Likewise, in order to constructively channel our pleasurable emotions, we need to find a worthy challenge well matched to our energies. So if we're feeling confident because we've just won a prize, rather than open the champagne, now's the time to boldly invite an exciting new partnership so as to build upon our achievements.

Pleasure and pain are natural allies and necessary opposites

Pain and pleasure are the reset switch for each other. One makes us keenly responsive to the other. If we're in pain, we hunger for pleasure, and if we're experiencing pleasure, a little pain is keenly felt and very motivating. They reactivate each other like sweet and sour, hot and cold, or night and day, just as the seasons make us grateful for the changes they bring. Progress is a process requiring both types of stepping stone – pain and pleasure, left and right – and no single part of the process is the goal.

Of course, I'm by no means the first to notice this left/right relationship. Circa 500BC, Heraclitus was saying, 'All things come into being by the conflict of opposites.' And rather more recently, the poet William Blake wrote, 'Without Contraries is no progression. Attraction and Repulsion, Reason and Energy, Love and Hate, are necessary to Human existence' (Blake, 1790).

If well-channelled pain and well-channelled pleasure lead to progress, what does this mean for our approach to everyday life?

Happiness should be what we feel when we make real-life progress, just as pain should be what we feel when thwarted. Therefore we should think carefully before we do anything to interfere with these energy-bringing and life-guiding messages. In other words we should not allow pharmaceutical or other contrivances to distort how our life feels. Instead, we might better consider our response to pain, because an occupational hazard of attempting to make progress is occasional pain. The more confident we can become in channelling our pains, the more confidently we can confront our fears and dare to venture forward despite the risk of increased setbacks or loss or rejection. By being more entrepreneurial in this way, we live life more openly and more deeply, and this brings the likelihood of greater progress, perhaps through greater love and knowledge.

Finally, it is important not to overlook the fact that progress itself produces a profoundly welcome physical and emotional feeling. It's no accident that we readily speak of a *sense* of progress, *pleasing* progress and a *desire* for progress.

How might the study of wellbeing help emerging technologies enhance rather than hinder our healthy relationships?

As well as its wonderful benefits, there are many ways in which ineptly used and overused technology weakens the healthy glue of our essential relationships. Through its all-pervasive media messages, its overload of information and choice, its excessive demands for communication and travel, the exponential growth in technology is exhausting us and diluting our vital relationships. The study of wellbeing has a prime role here in remedying these imbalances between our stock of personal resources and technology's increasing demands. We can, for example:

- guide technology producers to develop more 'relationship-friendly' products and services; and
- help people develop everyday strategies to better harness the advantages of technology to foster rather than hinder our fundamental human bonds.

Professor Lord Alec Broers was right to title his magnificent Reith Lectures of 2005 as *The Triumph of Technology*, but we should not allow technology to triumph at the expense of mankind.

CONCLUSION

I believe we should live life rather differently than we're doing at the moment, that we ought to dramatically shift our balance from me to we, from consuming to creating, from passive voyeur to active participant, from stagnant pleasure to relationship progress. To achieve this, our culture, our technological advances and the emerging science of wellbeing need to progress hand in hand with far more consideration for each other. We will all be the beneficiaries of such a healthy partnership.

Part VIII

Fairer?

22

Enhancement and Fairness

Julian Savulescu

One of the most commonly expressed objections to enhancement is on grounds of fairness. Opponents believe that allowing human enhancement will lead inevitably to unfairness, inequality and injustice:

> *'Improved' post-humans would inevitably come to view the 'naturals' as inferior, as a subspecies of humans suitable for exploitation, slavery or even extermination. Ultimately, it is this prospect of what can be termed 'genetic genocide' that makes cloning combined with genetic engineering a potential weapon of mass destruction, and the biologist who would attempt it a potential bioterrorist.* (George Annas, 2002)

> *The first victim of transhumanism might be equality. ... Underlying this idea of the equality of rights is the belief that we all possess a human essence that dwarfs manifest differences in skin colour, beauty and even intelligence. This essence, and the view that individuals therefore have inherent value, is at the heart of political liberalism. But modifying that essence is the core of the transhumanist project. If we start transforming ourselves into something superior, what rights will these enhanced creatures claim, and what rights will they possess when compared to those left behind?*
>
> *If some move ahead, can anyone afford not to follow? These questions are troubling enough within rich, developed societies. Add in the implications for citizens of the worlds poorest countries for whom biotechnology's marvels likely will be out of reach and the threat to the idea of equality becomes even more menacing.* (Francis Fukuyama, 2004)

These would be mere consumer decisions – but that also means that they would benefit the rich far more than the poor. They would take the gap in power, wealth and education that currently divides both our society and the world at large, and write that division into our very biology. (Bill McKibben, 2003)

Yet there is nothing new about enhancement. The wealthy have always had better access to the advantages of education, healthcare and technology. These already alter and enhance biology, albeit indirectly. The use of direct biological interventions does not raise any new ethical issues, but it does require us to think clearly about which theory of justice should govern society.

There are four different considerations which are relevant to deciding whether enhancement will result in unfairness or injustice:

1   the concepts of fairness or justice;
2   the concept and possibility of enhancement;
3   facts about the natural distribution of capabilities and disabilities; and
4   the concept of determinism.

FAIRNESS/JUSTICE

There are several theories of justice, including Utilitarianism, Egalitarianism, Rawls's Maximin (or Justice as Fairness) and Prioritarianism.

According to Utilitarians, enhancements should be distributed to provide the greatest benefits to the greatest number (in other words to bring about the most good). In this view, there is nothing unjust if enhancement is allowed and some people are worse off, even badly off, provided that enhancements are distributed according to a principle of equality which states that each is to count for one and none for more than one. That is, provided that enhancements are allocated strictly to bring about the greatest good, with no eye to social privilege, status, wealth or other irrelevant consideration, the resulting distribution is just.

According to Egalitarians, the basis for the distribution of enhancements should be to provide equal consideration of equal needs. Enhancements should alleviate need in individuals as much as possible – the greater a person's need, the greater that person's entitlement to resources.

According to Rawls's Justice as Fairness, we should distribute enhancements so that the worst off in society are as well off as they can be (Rawls, 1971). According to Prioritarians (Parfit, 1997), we should give priority to those who are worst off, but we should also aim to maximize wellbeing of all members of society.

A 'fair go'

In 'Rights, utility and universalization', John Mackie (1984) suggests everyone has a 'right to a fair go'. According to a maximizing version of giving people a 'fair go', we should give as many people as possible a decent (reasonable) chance of having a decent (good) life. This is a plausible common sense principle of justice and has been called 'sufficientarianism'.

According to the view of justice I favour, getting a 'fair go' is having a fair chance of receiving an intervention that has a reasonable chance of providing a reasonable extension of one's life and/or a reasonable improvement in its quality. A fair go entails that each person has a legitimate claim to medical care when that care provides that person with reasonable chance of reasonable extension of a reasonable life and/or a reasonable improvement in its quality. Comparable legitimate claims are those referring to similar needs, and as many comparable legitimate claims should be satisfied as possible. Provided as many comparable legitimate claims are being satisfied as possible, there should be equality of access.

Which view of justice we accept does not matter. For the purposes of argument, however, I will adopt here the right to a fair go.

ENHANCEMENT

Enhancement can be defined as something which makes our lives better. This is clearly desirable as an intrinsic good. What makes a good life is subject to debate, however: hedonists believe it is the pursuit of pleasure; others believe that it is found in desire fulfilment and others in the perfection of wellbeing.

There are a number of attributes that might characterize a good life, such as having good friends, knowledge, self-respect, autonomy or pleasure. Enhancement can help achieve the good life by providing instrumental goods: qualities that increase the chances of us having a good life, such as health or intelligence.

We can enhance our chances of leading a good life in four ways. We can manipulate our biology, our psychology, our social environment or the natural environment. Biological enhancement can improve our bodies, making us more beautiful, or stronger, or more intelligent, or stronger willed. Psychological enhancement could produce 'a better person' by psychological means. Social enhancement could improve our social environment, providing, for example, cleaner air, stable laws and a social security system. We can also intervene in our natural environment, for example by creating more parks, protecting nature or increasing sunlight. Not all forms of enhancement are controversial; not many would take issue with cleaner air or good schooling. There is controversy, however, surrounding biological and internal technological enhancements, and it is these that this chapter will now focus on.

Enhancement, disability and capability

The aim of any enhancement is to increase wellbeing, or to influence how well a life goes. Wellbeing depends largely on a person's capabilities and disabilities. Capability is a state of the person which increases the probability of achieving a good life. Disability is a state of the person which decreases the probability of achieving a good life. Disease is therefore a disability. To maximize how well people's lives go, therefore, we need to work to increase people's capabilities and reduce their disabilities.

What is a disability?

Disability typically falls under the following categories:

- deafness;
- blindness;
- paralysis; or
- cognitive impairment.

However, I have defined disability as a state which decreases the probability of achieving a good life. The capabilities that enable a person to live a good life vary depending on the environment. Therefore, what in one context might be called a disability might in other circumstances be a capability. An example of this is atopic tendency. Atopic tendency is a disability in the developed world, where it is more commonly known as asthma. However, in the developing world, it provides protection against worm infestation, and is therefore a capability. In the same way, deafness is not a disability in a very noisy world, but it is in our world.

We can decide that x is a disability in circumstances e if:

x reduces the chances of a person realizing a possible good life in circumstances e.

In order to decide whether a state is a disability or a capability, we need to fix or predict the social and other environmental circumstances prevailing.

Biology/psychology as capability/disability

Our biology contributes not only to our health, but also to how well our life is likely to go. All our biological or psychological states can influence our wellbeing in different environments and in the face of different challenges. We are all disabled in the sense that some features of our biology or psychology make it more difficult for us to live well. And it is possible to predict which attributes will be capabilities or disabilities in likely future environments.

Example: Self-control

In the 1960s Walter Mischel conducted impulse control experiments where four-year-old children were left in a room with one marshmallow, after being told that if they did not eat the marshmallow, they could later have two. Some children would eat it as soon as the researcher left. Others would use a variety of strategies to help control their behaviour and ignore the temptation of the single marshmallow. A decade later, it was found that those who were better at delaying gratification had more friends, better academic performance and more motivation to succeed. Whether the child had grabbed the marshmallow had a much stronger bearing on their SAT scores than did their IQ (Savulescu, 2006). Impulse control has also been linked to socioeconomic control and avoiding conflict with the law. Thus poor impulse control is a disability.

Other categories

Some qualities are always capabilities, or rather they are so in most likely environments or circumstances. In *From Chance to Choice*, Buchanan et al (2000) call these 'all purpose goods'. All-purpose goods include:

- intelligence;
- memory;
- self-discipline;
- foresight;
- patience;
- sense of humour; and
- optimism.

We can also identify other categories of traits which are generally beneficial:

1   Autonomy-enhancing traits: These are psychological capacities necessary for autonomy, including:
    - concept of self;
    - ability to remember, understand and deliberate on relevant information;
    - strength of will;
    - foresight; and
    - empathy.
2   Social traits: The ability to make friends, form close personal relationships, understand others' feelings and respond (it is this capability that psychopaths lack).
3   Our moral character: The capacity for virtues such as empathy, imagination, sympathy, fairness and honesty.

Other specific examples

Some traits are not so easy to categorize, however. Religiosity can be a capability, leading to a good life through trust in providence, and may aid endurance in suffering, as well as encouraging moral capabilities such as honesty and fairness. On the flipside, it could be seen as a disability, inducing feelings of guilt or leading to decision-making others would consider disabling, for example the refusal of medical treatment for religious reasons.

DISTRIBUTION OF CAPABILITIES AND DISABILITIES

Capabilities and disabilities are not distributed equally

The distribution of capabilities and disabilities is already uneven, through natural and social distribution. The 'normal distribution' of IQ and other biological capabilities and disabilities through genetic inheritance covers a wide range, and it is important to recognize that nature allots capabilities with no eye to fairness. Capabilities and disabilities are also distributed socially: self- or other-inflicted injuries, institutions, culture and education all advantage some and disadvantage others.

What does fairness require?

For a fair society, as many people as possible should be given a decent chance of a decent life. We already try to use social determinants of a good life, such as laws, attitudes and practices to ensure that women, different races and disabled people are given a 'fair go'. By manipulating the biological determinants of the good life, we could ensure that people have the capacity to have a good life in the way the world is likely to be in the future.

The example of IQ

IQ is distributed according to the famous bell curve. Low intelligence becomes a disease when IQ is less than 70. Intellectual disability is arbitrarily defined as an IQ less than 70 because IQs less than that, although within the normal distribution, severely affect people's chances, historically, of leading a good life. Medical research is directed at treating intellectual disability through biological or pharmacological means, and social resources are provided to support people with an IQ of 70 or less.

But what IQ is necessary for a decent chance of a decent life? Jim Watson once stated at a conference that he was most concerned about the bottom third and thought that they should be offered enhancements. Perhaps, in a technologically

Number of scores

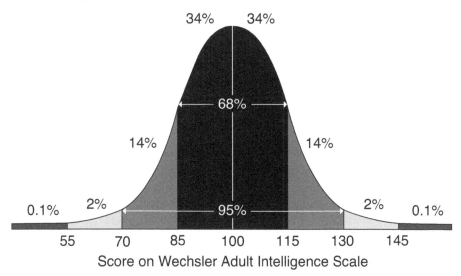

Figure 22.1 *Distribution of IQ scores*

Source: Encarta (2006)

sophisticated society, people would significantly benefit from a higher IQ. An IQ of around 120 is needed to be able to complete tertiary education, for example.

Justice/fairness requires we get as many people as possible up to the minimum IQ necessary for a decent chance of a decent life. Fairness thus requires enhancement. Far from being opposed to enhancement, justice requires enhancement. It is on this basis that we currently choose to treat those with an IQ less than 70. But where we set the minimum threshold for treatment or enhancement is up to us. It depends on how we define a decent chance of a decent life in the way society and the world are likely to be.

This is a general point. The need to provide as many people as possible with a decent chance of a decent life doesn't apply only to the enhancement of IQ. Any biological property which can constitute a capability or disability is suitable for enhancement. This includes:

- cognitive abilities;
- social abilities;
- impulse control; and
- physical abilities.

Insofar as these contribute to a reasonable chance of leading a reasonably good life, they are candidates for enhancement on the grounds of justice.

DETERMINISM

Causal determinism states that the past fully determines the present, which in turn fully determines the future. Complete knowledge of the state of the world and all the physical laws which govern it would allow us to predict every event in the future. Some have argued that causal determinism is incompatible with the concept of free will.

Other determinisms that have been identified include:

- *Genetic determinism* states that genes determine who we are. According to some genetic determinists, we should not create identical twins; since they will have identical genomes, they will be identical people with no individuality or autonomy. This is clearly false – identical twins are different people with autonomy. We can conclude that genes do not determine who we will be – they only contribute probabilistically.
- According to *technodeterminism*, technology will determine how our lives go. This is also false – there is a probabilistic relationship between technology and the future of society.
- According to *social determinism*, certain social structures and arrangements will inevitably determine other social structures and arrangements. This assumes people will necessarily act in a certain way. Social determinists believe that we will inevitably respond to new technologies in damaging ways.

An example of social determinism: Clonism

According to the so-called 'living in the shadow' objection to cloning, clones will have bad lives because they will live in the shadow of expectation and be the object of control, disapprobation and so on:

> *We should not create identical twins. Since parents will dress them the same and treat them as if they were the same person, they will have no individuality or autonomy.* (Holm, 1998)

This position denies free will and responsibility. Our justice system, on the other hand, assumes that we have the choice to act alternatively, and that as a result we are responsible for our choices. We do have power over how we treat people and new technologies. The response to the problem of 'clones and the shadow of expectation', therefore, is not to ban cloning but to remove the shadow of expectation. Likewise, we can decide not to dress and treat identical twins the same way in order to address problems with individuality.

To claim that clones will be treated adversely is a crude form of social determinism, which I have called 'clonism'. This is like racism, which assumes that blacks will be treated worse than whites. How we treat blacks, clones, IVF babies, people with two heads or dead people, however, is all up to us.

Social determinism, fairness and enhancement

McKibben (2003) argues that enhancement should not be permitted because it will be used to increase inequality: enhancement technologies will be available primarily to the wealthy, and so the rich get richer, the rich get smarter, the smarter get smarter and the smarter get richer.

However, this is not necessary and assumes social determinism. It is possible to use enhancement to increase inequality, but it is not a necessary outcome of enhancement. We have free will and a sense of responsibility: it is up to us how we choose to use this technology. We can use enhancement to increase inequality or to reduce it. If, for example, we raised the bottom third, as Jim Watson suggested, this enhancement would reduce not increase inequality.

An example of performance enhancement in sport: EPO

EPO is a natural hormone produced by the kidney which stimulates red blood cell production, thereby increasing the capacity of blood to carry oxygen. EPO is considered safe to a level that increases the red blood cells by 50 per cent, although through natural inequality, 5 per cent of the population have a natural level higher than this. The starkest example of naturally high levels of EPO is the Finnish skier Eero Mäntyranta, who won three Olympic gold medals in 1960–1964. It was subsequently found that he had a genetic mutation that meant that he 'naturally' had 40–50 per cent more red blood cells than the average person.

This natural inequality could be corrected by increasing red blood cell levels in those with 'normal' levels. Increases in EPO levels happen naturally as a result of pregnancy, haemorrhage and disease and can be increased through altitude training, autoinfusion, hypoxic air machines and injection of EPO. Alternatively, the red blood cell level of those with exceptionally high counts could be reduced to a 'normal' level by blood donation.

Thus we could easily and efficiently set a red blood cell level for competition to maintain and balance athlete safety and performance and the concerns of sport (for example to improve the spectacle for spectators). However, in sport we want to reward those who are naturally best. Sport is a celebration of those who have won life's lottery, and there is only one winner. But there is no reason why there has to be a person who comes last in life. If the unit were not red blood cells, but units of good life, is it really just and fair that there is a natural distribution in how well our lives go?

Social Not Biological Enhancement?

We accept social enhancement, and there are good reasons to prefer social to biological intervention, for example:

- if it is safer;
- if it is more likely to be successful;
- if justice requires it (based on the limitations of resources); or
- if there are benefits to others or less harm.

When there are good reasons to prefer social intervention to direct biological or psychological intervention, then we can say that disability is 'socially constructed' – it is caused or upheld by social structures or beliefs. However, the range of disabilities includes biological and psychological elements as well as socially constructed elements. It is possible to remove or alleviate disability at all levels, and we must consider reasons for and against biological and psychological intervention as well as the more commonly acceptable social interventions.

RESPONSES TO BIOCONSERVATIVES

We have seen that Nature allots advantages and disadvantages with no regard to fairness, and that enhancement can improve people's lives. The best way to protect the disadvantaged from the inequalities that bioconservatives like McKibben believe will follow from enhancement is not to prevent enhancement, but to ensure that the social institutions we use to distribute enhancement technologies work to protect the least well off and to provide everyone with a fair go. People have disease and disability, and egalitarian social institutions and laws against discrimination are designed to make sure that everyone, regardless of natural inequality, has a decent chance of a decent life. Enhancement to remove disability and disease would thus be an effective way of achieving a laudable aim.

CONCLUSION

Fairness requires enhancement. Which enhancements we adopt and how depends on which theory of justice we subscribe to, but failure to enhance may result in significant injustices. It is within our power to use technology and enhancement to bring about a more just society, where everyone has a fair go. Conservatives who oppose the use of biological and internal technological enhancements are guilty of a crude form of social determinism, predicting adverse social consequences of allowing enhancement and ignoring the fact that it is within our power to prevent these and reduce inequality.

We can allow our biology to be determined by the natural lottery or by wealth. Both of these lead to injustice. Judicious use of enhancement, based on a value-informed policy, on the other hand, can ensure that each of us, regardless of genetic or financial inheritance, has a 'fair go'.

23

Towards a Fairer Distribution of Technology in Maintaining Human Health: An Example of Child Immunization in Western China

Zhao Yandong and Ma Ying

The People's Republic of China started its national programme for immunization of children in the 1970s. At present, vaccination coverage has been steady at around 85 per cent for some years. However, there are signs of unequal distribution of vaccination.

Despite the fact that vaccination is the cheapest and most common technology to prevent infectious diseases, it is not available to all the people who need it due to factors related to income, knowledge and medical services, amongst others. Consequently, infectious disease is still one of the biggest threats to people's lives. According to statistics from the Ministry of Health in China, the death rate from tuberculosis jumped to the highest among all infectious diseases by 2000. Western provinces, rural areas, the poor and children are facing the greatest challenges in fighting against tuberculosis.[1]

In this chapter we discuss some of the inequities in child immunization in China and consider some of the factors influencing them.

IMMUNIZATION OF CHILDREN IN CHINA

Under years of development, China has set up a vertically administered immunization system. The health administration under the State Council is the highest-level administrator of child immunization work. It is responsible for setting down the national immunization programme. Based on the national programme, the provincial health administrations make their own programmes that are adapted to their particular situation. The health administrations at the county level assign qualified healthcare institutes to perform vaccinations. BCG (the standard vaccine against tuberculosis) and the first Hepatitis B injection are usually carried out in the hospital where the child is born. Later vaccinations are carried out in township/commune health centres in rural areas or community health centres or hospitals in urban areas. Vaccination can also be conducted by qualified personnel (for example a mobile medical team) at places other than the health centres on special occasions.

According to the national immunization programme, every child should get one BCG, three polio, three DPT (combined vaccines for diphtheria, whooping cough and tetanus), one measles and three Hepatitis B vaccinations within the first year after birth. Hepatitis B was included in the national immunization programme in 2002.

The Chinese government has made great efforts to provide free vaccines to children. However, due to the lack of funding in local health institutes, a labour or injection fee is charged in many places. The newly implemented Regulation on Vaccine Transportation and Vaccination (2005) provided the highest and most complete legal protection to immunization work in China. The regulation states that every child vaccination included in the national immunization programme is free of *any* charge. Funding is provided by government revenues.

As a result of these efforts, child immunization rates are increasing steadily in China. According to the Ministry of Health, China achieved the goal of a 85 per cent coverage rate for child immunization (BCG, polio, DPT and measles) at the provincial level in 1989. In 1991 and 1995, the 85 per cent coverage rate was also achieved at the county level and commune level, respectively.

Despite these achievements, however, China's immunization programme still faces some problems and challenges. Since the 1980s, the vaccination rate of the four vaccines (BCG, polio, DPT and measles) for children under one year old has been steady, at around 85 per cent. Some places have even showed signs of decrease in the vaccination rate (Chinese Ministry of Health, 2005). In addition, the vaccination rate varies by region and social group. Remote areas, the poor and migrant people are the weakest links in immunization work.

To solve these problems, in-depth analysis on the distribution of vaccination technology in China is needed. This chapter examines vaccinations on children under five in Western China, based on empirical social survey data, with the aims of deepening our understanding of child immunization in Western China and providing policy suggestions for decision-makers.

DATA

The main data for this chapter come from the MEDOW[2] survey of 11 provinces in Western China. The survey was carried out by the National Research Centre for Science and Technology for Development, Beijing, and the Fafo Institute, Oslo, and financed by the Norwegian Ministry of Foreign Affairs. Fieldwork was carried out in 2004 and the survey sampled 44,000 urban and rural households. The survey collected information on almost all aspects of people's living condition, including housing and infrastructure, education, labour force, health, immigration, agricultural activity, household economy, childbirth and immunization. The survey also collected information on the community where the survey was conducted. A separate questionnaire on vaccination was administered for all the children under five living in the interviewed household, with the child's carers as the respondents.

WHO DID NOT GET IMMUNIZATION? AND WHY?

Our survey data showed that the vaccination rate of children under five years of age in Western China is 90.4 per cent.[3] This meant that 9.6 per cent of children under five, or about 1.6 million children, had never been vaccinated in the 11 western provinces. Most of these unvaccinated children lived in rural areas and in households with lower income and less education.

To better understand the mechanism of the distribution of vaccination, we need a closer examination of who did not get vaccination and why. We summarized the factors influencing the distribution of medical technology among the public into three main categories, or 'three A's':

1   affordability;
2   accessibility; and
3   acceptability.

In the following, we will discuss these three categories and their relationship as well as focusing on a number of geographic and demographic background variables.

AFFORDABILITY

The core question of affordability is whether people with different economic status, especially those with lower income, could enjoy the benefit brought by the progress of technology. In the study of equality as related to science and technology, affordability is one of the issues that first drew people's attention and is also still the most frequently discussed. Many studies have proved that income is one of the most important factors determining people's utilization of medical technology (Zheng et al, 2005). How to help those vulnerable groups who were not able to afford the needed medical technology has become the key issue of health policy in many countries.

In our study, affordability is measured by income and the subject's self-assessment of household economy. Using both objective and subjective indices of income provides us with a more comprehensive measurement of people's economic status that using income alone.

We divided the households into five groups according to their income, and found that the proportion of children without any vaccination rose rapidly as the income dropped. Only 1.4 per cent of children in the 20 per cent of households with the highest income never had any vaccination, while the percentage of unvaccinated children in the 20 per cent of households with the lowest income was as high as 18.2 per cent. A similar correlation was found between child vaccination rate and self-assessed household economic status. In the groups who assessed their household economy as 'good' and 'fairly good', only 4.9 per cent and 3.6 per cent of children never had any vaccination. This rate leaped to 17.1 per cent in the group of households who assessed their economic status as 'poor'.

This corresponds to other results we got from the questionnaire. About one third of carers (the largest portion) choose 'No money/too expensive' as the reason for not getting vaccination.

ACCESSIBILITY

Accessibility relates to the extent to which different social groups have equal access to the basic services and infrastructures of technology. Some researchers have pointed to this as one of the main factors in securing high vaccination rates (Zhu et al, 1998). Other researchers have even argued that the impact of income on health might have been overestimated, and that the main factor influencing health is access to public health facilities (Anand and Ravallion, 1993; Wang et al, 2002).

We measured accessibility to vaccination in two ways. First, we asked the community leader where children in the community usually get their vaccination. The answers were categorized as:

1   'A hospital/clinic in this village/neighbourhood';
2   'A hospital/clinic in the commune'; and
3   'By mobile medical team/other place'.

We assume those who selected the first option had better access to vaccination. Second, we asked whether the child was born in a hospital/clinic or at home, assuming those born in a hospital/clinic would have better access.

The results showed that accessibility strongly influenced children's vaccination take-up. Only 7.6 per cent of children in the community with a hospital or clinic did not get vaccination, while the proportion of unvaccinated children jumped in the communities where children usually get their vaccination in the commune (11.7%) and other places (12%).

Children born in a hospital/clinic have a much lower unvaccinated rate (2.5 per cent) than those born at home (17.5 per cent).

ACCEPTABILITY

Beyond accessibility and affordability, there is yet another important aspect influencing the distribution of medical technology. Even if a technology is accessible and affordable, if people do not have the right knowledge or do not understand the value and importance of the technology, they might have less chance of benefiting from it. This aspect was termed as 'acceptability' in this study. As indicated in many other studies on equality in science and technology, one of the most efficient ways to promote people's acceptance ability of new technology is to increase their educational level. Studies have proved that increasing a mother's level of education significantly improves children's health (He et al, 2002). In our case of vaccination, we hypothesized that the mother's education level would affect children's probability of getting vaccinated.

This hypothesis was supported by our data. As many as 21.7 per cent of the children whose mother 'never attended school' did not receive the vaccination, while the number dropped to 12.9 per cent in the group of children whose mother had 'finished primary school'. The rate of non-vaccination kept dropping with the increase in the mother's education level. In the group of children whose mother's education level is 'higher than college', only 1 per cent of children did not get vaccinated.

Another result supported our hypothesis. We asked the community leader if the community had held any health education programme – promoting knowledge of health and sanitation – in the last year. In the communities that didn't implement such programmes, the percentage of unvaccinated children was as high as 15.9 per cent, whereas in those communities that implemented them, the rate of un-vaccinated children dropped to 7.1 per cent. Popular dissemination of relevant knowledge apparently improved the acceptability of vaccination.

GEOGRAPHIC AND DEMOGRAPHIC VARIATIONS

We also explored the vaccination rate in different geographically located communities and demographic groups. A much higher proportion of children living in rural areas (11.0 per cent) were not vaccinated than their counterparts in urban areas (3.2 per cent). This result is not surprising given the huge rural–urban divide in China. The proportion of unvaccinated children also varied a lot among communities located in flat areas (4.7 per cent), hilly areas (9.0 per cent) and mountainous areas (16.5 per cent). As for age groups, children under one year old had the highest proportion of unvaccinated population (16.9 per cent), while in other age groups – from one to four years of age – the proportion was stable, at around 7–9 per cent.

MULTIVARIATE ANALYSIS

To estimate the relative effects of the factors mentioned above, we conducted a multivariate logistic regression analysis, taking 'whether the child had any vaccination' as the dependent variable, and the aforementioned variables as independent variables. We included other variables, such as urban–rural, age, sex and geographical location of the community, as the control variables. As the differences among provinces are quite large, we also included provinces as control variables.

The analysis shows that, of all the independent variables in the equation, 'born in a hospital/clinic' and 'mother's education' are the most influential, showing that accessibility and acceptability are very important for children's vaccination. 'Income' and 'self-assessment of economy' are also significant, but less important than accessibility and acceptability.

Among the control variables, two in particular should be mentioned here:

1   The urban–rural difference is very significant: children living in urban areas have a higher chance of vaccination.
2   If the community implemented any public sanitation programme in the last year, this will greatly increase the children's probability of vaccination.

CONCLUSION AND DISCUSSION

A fairer distribution of technology in maintaining human health is a target that government and society are pursuing. This chapter, taking child immunization as an example, argues that this target could be operationalized into ensuring people's affordability, accessibility and acceptability of (new) technology. Our analysis showed that mother's education is the most important factor influencing

Table 23.1 *Log regression of likelihood of child vaccination (not vaccinated = 1)*

Independent variables	B	S.E.	Exp(B)
Affordability			
Household income per capita (logged)	−0.15**	0.06	0.86
Self-assessment of household economy	−0.20***	0.07	0.82
Accessibility: location where most children in community get vaccinated (reference = other place/mobile team)			
Vaccinated in this community	−0.56***	0.13	0.57
Vaccinated in this commune	−0.09	0.15	0.91
Birth place of the child (reference = at home/ other place)			
In hospital/clinic	−1.32***	0.15	0.27
Acceptance ability			
Education of mother (years)	−0.14***	0.02	0.87
Have health education programme in the community (yes = 1)	−0.48***	0.11	0.62
Control variables			
Hukou registration (rural = 1)	0.62**	0.22	1.86
Sex (female = 1)	0.20	0.11	1.22
Age of child	−0.41***	0.04	0.66
Location of community (reference=flat areas)			
Hilly areas	−0.24	0.19	0.79
Mountainous areas	0.12	0.13	1.13
Provinces (reference = Neimenggu)			
Gansu	1.10	0.56	3.01
Guangxi	1.22	0.54	3.40
Guizhou	1.30	0.53	3.68
Chongqing	1.05	0.58	2.86
Ningxia	0.85	0.54	2.34
Qinhai	1.16	0.54	3.20
Shangxi	1.19	0.61	3.27
Sichuan	2.17	0.54	8.74
Xingjiang	0.91	0.55	2.48
Yunnan	1.46	0.53	4.30
−2 Log Likelihood	2533.392		
Cox and Snell R Square	0.11		

children's immunization, indicating that acceptability is most important for the fairer distribution of technology. This result has, at the same time, echoed with findings of other researches in terms of the value of increasing the education level of rural women. Another interesting finding in our study was that health education programmes had significantly promoted the children's likelihood of vaccination. This implied that the government could and should play a more active role in increasing people's acceptability of technology, for example by providing more health-related knowledge to a wider range of the population.

Our data also showed that accessibility is more important than income in ensuring equality in technology and health. Thus the government needs to provide more and higher-quality public health infrastructures and services, with a special concern for vulnerable groups, so as to maintain 'equality from scratch' in society. As indicated in other findings in our project, rural areas, mountainous areas and the areas where migrants are concentrated are in urgent need of more accessible healthcare facilities. These should become the main concerns of the government in the future.

Although in our study income is less influential than the two other aspects, it is still quite significant. Thus strengthening people's ability to pay is still an indispensable component in the government's effort to ensure equality in medical technology. This could be done in two ways. One is to promote economic development and to heighten people's income level. The other is to lower the price threshold of provision for vulnerable groups, for example by providing subsidies for poor and isolated populations or free vaccination for them. The Regulation on Vaccine Transportation and Vaccination implemented in 2005, declaring free vaccination for all children, showed the Chinese government's determination to achieve a fair distribution of medical technology amongst the public.

NOTES

1   See www.fh21.com.cn/yl/tebie/jiehe/jiehe.htm, accessed 23 May 2008 (in Chinese).
2   MEDOW (Monitoring Social and Economic Development of Western China) is a large-scale household survey of living conditions. The survey covers 44,738 households in 11 provinces of Western China.
3   Note that the vaccination rate here is calculated as the proportion of children under five who had received any vaccine. This is not directly comparable to the national statistics.

24

Ableism, Enhancement Medicine and the Techno-Poor Disabled

Gregor Wolbring

Advances in and convergence of:

1 nanoscience and nanotechnology,
2 biotechnology and biomedicine,
3 information technology,
4 cognitive science, and
5 synthetic biology (which can be described as the design and construction of new biological parts, devices and systems, and the redesign of existing, natural biological systems for useful purposes; Wolbring, 2006a),

otherwise known as 'NBICS' (for nano-bio-info-cogno-synbio), are envisioned to lead to applications in areas such as the environment, energy, water, weapons and other military applications, globalization, agriculture, security (Institute of Nanotechnology, 2005; Kostoff et al, 2006) and the global problems of disease and ill health (Roco and Bainbridge, 2002). Others perceive NBICS as an enabler for the pursuit of extreme lifespan extension, if not immortality, in other words to 'defeat the disease death', and of 'morphological freedom' (Sandberg, 2001).

Indeed, one of the most consequential advances in NBICS is the increasing ability, demand for and acceptance of changing, improving and modifying the human body and allowing it to move beyond species-typical functioning, perceiving the species-typical body as defective, diseased and in need of 'therapeutic enhancements'. On the level of non-humans, the generation of new life-forms (synthetic biology) is pursued, enhancing biological organisms in terms of their structure, function or capabilities beyond their species-typical boundaries.

Many different forms of enhancement are proposed, with many different purposes. Each form and purpose of enhancement comes with its own sales pitches, social consequences, problems and implications (Wolbring, 2005). Some are focusing on lower life-forms (Wolbring, 2007a), some on animals (Wolbring, 2007b). In this chapter, I cover the dynamic of the *Homo sapiens* enhancement discourse (Wolbring, 2005). One of the main arguments in the enhancement debate is that you can and should make a distinction between therapy and enhancement, or therapeutic versus non-therapeutic enhancement (Wolbring, 2005). However, the tenability and usefulness of such a distinction and of many other arguments employed in the enhancement debate depend on what concepts of health and ableism one follows, recognizing that both concepts influence each other.

Many of the enhancements are pursued within a pathology language. This chapter characterizes today's ableism and health concept discourses and illustrates the difficulties they pose for most arguments employed to demand a prohibition of enhancements in general or therapeutic enhancement in particular. It presents some of the problems and policy implications which come with enhancements and the shift in the health and ableism concept discourse and draws some conclusions.

ABLEISM

Ableism is a set of beliefs, processes and practices that, based on abilities one exhibits or cherishes, produce a particular understanding of oneself, one's body and one's relationship with others of one's species, other species and one's environment, and that understanding includes that one is judged by others (Wolbring, 2007c, d, e). Ableism exhibits a favouritism for certain abilities over others projected as essential while labelling real or perceived deviations from or lack of these 'essential' abilities as a diminished state of being, leading or contributing to justifying various other 'isms' (Wolbring, 2007c, d, e).

Ableism is an enabling umbrella ism for other isms such as racism, sexism, castism, ageism, speciesism, anti-environmentalism, GDP-ism and consumerism. In other words various isms can be seen to converge under the umbrella of ableism. However, one can also identify many different forms of ableism (A), such as:

- biological structure-based ableism (B);
- ableism inherent to a given economic system (E);
- cognition-based ableism (C); and
- social structure-based ableism (S).

Thus ABECS could be used as the ableism equivalent to the NBICS science and technology (S&T) convergence.

Ableism and favouritism of abilities is rampant today and has been throughout history. Ableism shaped and continues to shape areas such as human security

(Wolbring, 2006b); social cohesion (Wolbring, 2007f); social policies; and relationships among social groups, individuals and countries, humans and non-humans, and humans and their environment (Wolbring 2007c, d, e). Ableism is one of the most societally entrenched and accepted isms.

Ableism and S&T

The direction and governance of S&T and different forms of ableism have always been interrelated, and ableism will become more prevalent and new forms of ableism will appear with the increasing ability of new and emerging S&T:

- to generate human bodily enhancements in many shapes and forms, with an accompanying ability divide and the appearance of the external and internal techno-poor disabled (Wolbring, 2006b);
- to generate organisms or genomes from scratch (synthetic biology);
- to modify and ability enhance non-human life-forms (from microbes to animals);
- to separate cognitive functioning from the human body;
- to modify humans to deal with the aftermath of anti-environmentalism; and
- to generate products atom by atom, moving trade from nature-based commodities towards atomic-generated commodities. Already nano-modified commodities like rubber are much more durable, leading to a change in demand for natural rubber, which impacts onto natural rubber trade and workers.

We can already observe a changing perception of ourselves, our bodies, and our relationships with others of our species, other species and our environment. New forms of ableism (for example transhumanization of ableism) (Wolbring, 2006b, 2007c, d and e) are often presented as a solution to the consequences of other ableism-based isms such as speciesism (Institute of Nanotechnology, 2005) *and* anti-environmentalism.

Ableism against the traditional disabled people[1]

This form of ableism reflects the obsession with species-typical normative abilities, leading to discrimination against those perceived as 'less able' or 'impaired'. It is supported by the medical, deficiency and impairment categorization of disabled people (the 'medical model'). The medical model of disability (Wolbring, 2007h) perceives people to be disabled by their sub-species-typical functioning and rejects a social model of disability (Wolbring, 2007h) which sees people as disabled by the attitudinal and environmental barriers they face due to their body structure or function. It supports the deficiency categorization of 'sub-normative people'. It rejects the 'variation of being', biodiversity notion of these people (the 'social model'). It leads to the focus on fixing the person or preventing more such people being born, under the label of 'therapeutic interventions', and ignores the acceptance and accommodation of such people in their variation of being (Wolbring, 2005) .

This form of Ableism depends, among other things, on the medical model of health.

THE MEDICAL MODEL OF HEALTH[2]

Within the medical model of health and disease, health is limited to 'medical health' and is characterized as the normative functioning of biological systems, while disease or illness is defined as the sub-normative functioning of biological systems. Its method for locating the cause of and solution for 'ill medical health' comes in two forms:

1   Medical determinants of the medical health of a patient place the cause of sub-normative functioning within the individual patient's biological system, leading to medical interventions towards the species-typical norm. At the level of the individual this focuses on medical cure, medical individualistic care and individualistic normative rehabilitation as the primary end-point; at the political level the principal response is to make curative medicine more efficient (Wolbring, 2005). These are medical, individualistic cures (Wolbring, 2004, 2007h).
2   Social determinants/interventions of medical health of a patient, which focus on external factors such as contaminated water leading to bacterial or parasitic infections or job insecurity contributing to stress and heart disease (Wolbring, 2005, 2007h).

The main arguments in the enhancement debate that one can and should make a distinction between therapy and enhancement or therapeutic and non-therapeutic enhancement depend on the medical model of health, a medical model understanding of disabled people as impaired and the classic form of ableism. However that is history, and a third-generation model of health and understanding of disabled people has already appeared (Wolbring, 2005, 2006d) (the second-generation social model (Wolbring, 2005) is not covered in this chapter), with an accompanying change in the meaning of ableism (Wolbring, 2005, 2007c, d, e, g, h).

THE TRANSHUMANIST MODEL OF HEALTH

Within the transhumanist model of health, all biological species bodies, *Homo sapiens* bodies included, no matter how conventionally 'medically healthy', are defined as limited and defective as in ill health, and in need of constant improvement made possible by new technologies appearing on the horizon (a little bit like the constant software upgrades we do on our computers). Health in this model is the concept of having, at any given time, obtained maximum enhancement (improvement) of one's abilities, functioning and body structure (Wolbring, 2005, 2006d, 2007h). Consequently, 'disease' comes to include any unenhanced or outdated body (Wolbring, 2005, 2006d, 2007h), making every person with a non-enhanced body a potential healthcare client. Within this model, a non-impaired person is one who obtains the expected bodily enhancement. Impaired people are those who are unable, or who do not want, to improve themselves beyond *Homo sapiens* normative functioning (the 'techno-poor disabled'; Wolbring 2006e).

Under this model, science and technology interventions which add new abilities to the human body are seen as the remedy for ill medical health. These interventions see the enhancement of the body beyond species-typical parameters as a therapeutic intervention (transhumanization of medicalization; Wolbring, 2005). Enhancement medicine is the new field providing the remedy through surgery, pharmaceuticals, implants and other means (Wolbring, 2005).

Linked to the transhumanization of health is the transhumanization of ableism.

THE TRANSHUMANIZATION OF ABLEISM (GENERIC FORM)[3]

The transhumanized form of ableism is a set of beliefs, processes and practices that perceive the improvement of biological structures beyond species-typical boundaries as essential. It perceives species-typical biological structures as deficient, as being in need of constant improvement and as a diminished state of being.

The transhumanization of ableism related to humans[4]

Until now, a non-impaired person has been seen as someone whose body functioning performs within *Homo sapiens*-typical parameters. This is changing, however. The ability of new and emerging science and technology products to modify the appearance of the human body and its functioning beyond existing normative species-typical boundaries allows for a redefinition of what it means to be non-impaired (Wolbring, 2005).

One transhumanized form of ableism is the set of beliefs, processes and practices that perceive the 'improvement' of human body abilities beyond typical *Homo sapiens* boundaries as essential. It exhibits the favouritism of beyond *Homo sapiens*-typical abilities and perceives human bodies as limited, defective and in need of constant improvement and as being in a diminished state of being human if they are not enhanced beyond *Homo sapiens*-typical abilities. In tune with the belief that the human body is deficient (the 'transhuman medical model'; see

Wolbring, 2005), we are moving towards changing the body to expand its abilities beyond those typical for *Homo sapiens* (the 'transhuman medical determinant'; see Wolbring, 2005). This transhumanized version of ableism elevates the existing medicalization dynamic (Wolbring, 2005) to its ultimate end-point, where one classifies enhancement of the body beyond species-typical abilities as a therapeutic intervention (the 'transhumanization of medicalization'; see Wolbring, 2005), making the therapy/enhancement division or therapeutic versus non-therapeutic enhancement division untenable.

THE CONSEQUENCES OF ENHANCEMENT AND THE TRANSHUMANIZED VERSIONS OF HEALTH AND ABLEISM

I don't believe that we can prevent enhancement technologies from developing under the given societal realities and adherence to certain forms of ableism. This poses some imminent problems, which current systems of governance for science and technology are unlikely to be able to deal with. The following sections summarize a few of the problems I foresee.

International document interpretation

To see the differences between the three models of health, and in particular the potential effect of the transhumanist model, one only has to look at these quotes from the Bangkok Charter for Health Promotion in a Globalized World (WHO, 2005) and the Universal Declaration of Human Rights (United Nations, 1948) and think what the different definitions of health would mean for the scope and actions required:

> *The United Nations recognize that the enjoyment of the highest attainable standard of health is one of the fundamental rights of every human being, without discrimination.*

> *Regulate and legislate to ensure a high level of protection from harm and enable equal opportunity for health and wellbeing for all people.*

> *Government and international bodies must act to close the gap in health between rich and poor.*

> *Everyone has the right to a standard of living adequate for their health.*

Personhood

All UN-based documents use the term 'person'. However the term is not set in stone. Throughout history, many human beings have not been seen as persons, and in some places some are still seen as non-persons today. How do we define human beings? Will we go beyond what can be defined scientifically as *Homo sapiens*? What are the criteria for personhood? Do we have to redefine personhood to take into account new technological realities? How does any given redefinition of personhood affect people perceived as persons today? Might some people who are perceived as persons today become non-persons? These are all questions human enhancement raises. Further, the proposed cognitive enhancement of animals (Wolbring, 2007b) as a means to fight speciesism is another facet of the personhood debate. Indeed if projects which try to generate cognitive abilities in animals (Wolbring, 2007b) and cognition as part of artificial intelligence come to pass, a variety of other questions have to be asked (see Wolbring, 2007i).

Ableism-driven speciesism[5]

Speciesism assigns different values and rights to beings based on their species. Human beings are seen as superior to other species due to their exhibition of 'superior cognitive abilities'. This speciesism has led to behaviours whereby human beings have dealt with other species according to 'we can do it, so we do it'.

The transhumanized version of ableism related to non-human species[6]

Another transhumanized version of ableism is the set of beliefs, processes and practices which champions the cognitive enhancement of animal species beyond species-typical boundaries, leading to cognitively or otherwise 'enabled species'. This is seen as a way to alter the relationship between human beings and other species, and to change how non-human species are judged and treated.

This is often the approach: instead of questioning the tenets of ableism, one tries to find ways for a disadvantaged group to become as able. 'I can be as able as you are, I am as able as you are' can be heard often.

This version of ableism favours cognitive abilities. There are other examples. Besides racism and speciesism, favouritism towards cognitive abilities might also play a role with respect to artificial intelligence, which may gain equal status to people when it is seen as cognitively able enough. Human rights might then become an obsolete concept, as rights might no longer be based on the fact of being human but on having a certain level of cognitive abilities (sentient rights). If it is possible to separate cognitive abilities and consciousness from the human biological body, the resulting entity would gain rights by itself – independent of the body.

The ability divide[7]

The more forms of enhancement become available, the bigger the ability divide will become. This would follow the pattern of the divides developed after the introduction of other technologies. As we seem not to be able to close any of the other divides (for example 98 per cent of webpages are still not accessible to blind people), it is doubtful we will be able – under current policies – to close the ability divide. Indeed people and groups who promote human enhancement use the existence of other societally accepted divides to further their cause. As the World Transhumanist Association states:

> *Rich parents send their kids to better schools and provide them with resources such as personal connections and information technology that may not be available to the less privileged. Such advantages lead to greater earnings later in life and serve to increase social inequalities.*
> (World Transhumanist Association, 2003)

A debate has to take place about which divides are acceptable under what conditions and why.

The techno-poor disabled and the ability divide[8]

Transhumanists and others propose that wealth will eventually trickle down (World Transhumanist Association, 2003). However, if this is the case, why do we still have poor people, unclean water, and many places without phones and electricity? Every technology has led to a new group of marginalized people and to new inequalities, and there is no reason under today's policy realities why this would be different if the human body becomes the newest frontier of commodification. As much as human enhancement technology will become an enabling technology for the few, it will become a disabling technology for the many. As more powerful, less invasive and more sophisticated enhancements become available, the market share and acceptance of enhancement products will grow in high-income countries. This will very likely develop into a situation where those who do not have or do not want certain enhancements (the techno-poor disabled) will be discriminated against, will be given negative labels and will suffer difficult consequences in line with how the 'traditional disabled' are treated today.

For any given enhancement product there will not be a bell curve distribution, but rather a distribution jump from the 'have nots' to the 'haves', which will lead directly to an ability divide. What will change – depending on social realities such as GDP, income levels and other parameters – is how many people end up as haves and have nots (the techno-poor disabled).

This ability divide will develop between the poor and rich within every country. And it will be bigger between low- and high-income countries than within any given country. Not everyone will be able to afford enhancement of their body, and no society will be able to afford to enhance everyone, even if they wished it. Thus billions of people that today are seen as healthy will become disabled, not because their bodies have changed, but because they have not changed their bodies in accordance with a transhumanist norm.

I believe that if we go down the road of enhancement, we need to change the whole system towards distributive justice, giving the enhancements first to those who need them most. And as this is not very likely to happen, the second best option is to absolutely ensure that no one can gain any positional advantage from enhancements and no one can force their desires and self-perception on others, whether it is their child or child to be or anyone else. If we go on as we are today, we will see the appearance of a new underclass of people – the unenhanced.

Responsibility

Enhancement will require changes to the concept of responsibility. The transhumanists consider it to be a parental responsibility to use genetic screening and therapeutic enhancements to ensure as 'healthy' a child as possible (World Transhumanist Association, 2003). Under such a model, would it be child abuse if parents refused to give their children cochlear implants, if they felt there was nothing wrong with their child using sign language, lip reading or other alternative modes of communication? Would it be child abuse to fail to provide a 'normal' child early in life with a brain–machine interface?

Increase in numbers of people perceived as impaired

Finally, as enhancement technologies are developed, the definition of being 'impaired' will change and the number of people so perceived will increase (Wolbring, 2005). The transhumanist model sees every human body as defective and in need of improvement, such that every unenhanced human being is, by definition, impaired.

It might be assumed that 'traditional disabled people' would welcome such a shift, as it would move the focus away from particular forms of impairment, towards the ability to enhance oneself – a challenge which 'traditional disabled people' would share with other 'unenhanced people'. Indeed many transhumanists are very aware of the potential to use disabled people as a trailblazer for the acceptance of transhumanist ideas and products (Wolbring, 2005). For example, James Hughes, the executive director of the World Transhumanist Association, has written, 'Although few disabled people and transhumanists realize it yet, we are allies in fighting for technological empowerment' (Hughes, 2004).

However, as many 'traditional disabled people' are poor and live in low-income countries, they have far more to lose than gain from such a shift. They might think that they are better off because they share their lack of ability with others who can't afford enhancement, but we can expect that resources would never be 'wasted' on people who are below the traditional norm. This is because with the same amount of money more people who already fit the traditional norm could be enhanced than people who are different.

As Murray[9] and Acharya have written, 'individuals prefer, after appropriate deliberation, to extend the life of healthy individuals rather than those in a health state worse than perfect health' (Murray and Acharya, 1997). What this means is that it is realistic to expect that if we follow the same model, decision-makers will choose to enhance the lives of healthy individuals rather than those in a state of less than perfect health, because this will be seen as better value for money.

The spirit of the above quote might lead to societal development that favours 'enhancement medicine' over 'curative medicine', seeing pure curative medicine as futile and a waste of healthcare dollars (Wolbring, 2005).

NOTES

1 See Wolbring (2006b, 2007c, d, e, g, h).
2 See Wolbring (2005, 2007b).
3 See Wolbring (2005, 2007c, d, e, g, h).
4 See Wolbring (2005, 2007c, d, e, g, h).
5 See Wolbring (2007d, g).
6 See Wolbring (2007d, g).
7 See Wolbring (2006b, e, 2007g).
8 See Wolbring (2006b, e, 2007g).
9 The originator of 'disability adjusted life years' – a measure developed to give decision-makers a tool to judge where money should go to in health interventions.

Part IX

Governable?

25

Governance of New and Emerging Science and Technology

Arie Rip

Are actual and projected scientific and technological developments governable? The answer to this question has to be an ambivalent 'yes and no'. If one thinks of governing as intentional and driven from a centre, the answer is no. But when one gives up on this illusion, there is a lot that can be done. This is what I will address in this chapter, also building on our experience with constructive technology assessment of nanotechnology (Rip et al, 1995; Schot and, Rip 1997; Rip, 2007).

Of course, new life sciences technologies can be stimulated or slowed down, they will be debated and shifts will occur. There is no autonomous development, and in that sense they are governable. However, they are not governable in any direction one would wish, and not from a central position. There are many hands involved, both visible and invisible (Rip and Groen, 2001), while outcomes are the often unintended effects, at a collective level, of interacting and contending actor strategies.

In such a situation, governance as control is possible only to a very limited extent. The trajectory of stem cell research in the US is a case in point. The Bush Administration has introduced strict regulation, but this only applies to federally funded research. Private funding is used to support a lot of stem cell research, and now some of the States are also investing. While the debate about Bush's strict regulation continues, stem cell research on the ground flourishes.

Governance as modulation of ongoing processes is a serious option, however. What happens is shaped by actor strategies, their interactions and emerging networks. These can be modified, and when recognized for what they are, intentionally modified. The latter can be called 'modulation'. Governance as modulation occurs anyhow, but may not be very productive when an illusion of control is the dominant driver. The recent interest in policy networks as a form of governance at an intermediary level is more realistic.

In the case of new and emerging science and technology, heterogeneous networks linking science, industry, policymakers and different societal stakeholders are important. These can be mapped and modified, and analysed as to how they enable and constrain future developments. Anticipation on such (in a sense endogenous) futures allows more intelligent, though not necessarily always more effective, modulation.

Taking the time dimension into account, one can speak of co-evolution of science, technology and society (Rip, 2006). This co-evolution is now becoming more reflexive (in the sense used by Beck (1992): effects of the development return, sometimes with a vengeance, to the door of the enactors of the development), which then leads to reflections and responses modifying the co-evolution. An example is the way nanotechnology enactors are concerned about a potential impasse, as happened with genetic modification technologies in the agrofood sector. They refer back to the impasse around GM food, and search for ways to avoid a similar one – including calls like 'this time, make everything right from the beginning'.

For human enhancement technologies, the enactors are carried away by the prospects, and only grudgingly recognize that there might be problems. If there are ethical dilemmas, the enactors feel they should be taken care of by others – a division of moral labour.

Turning now to the structure of these promises of enhancement, and what that implies for governance – the debate centres on the nature and desirability of enhancement. It misses out a key point, and one which might explain the broad reluctance to accept some forms of enhancement, such as learning pills for pupils, and much less such reluctance for others, like supporting their education. It is the fact that what is being offered are, as I will argue, upstream solutions for downstream problems. Actually, this point applies to almost all new and emerging science and technology in our society. The debate on human enhancement is then a case where the structure of such promises is particularly visible.

New options are positioned as solutions for problems downstream with respect to the situation where the intervention is to be applied. For example, if we can find a gene (or a cluster of genes) responsible for an undesirable trait (or a desirable one, for that matter), we propose to modify the gene, so that the eventual situation 'downstream' will be better. In fact, when there are new findings, there is a search for problems that they might help to solve. This runs from rhetorical claims when applying for funding and other resources to setting directions for R&D programmes, as happened with the so-called Green Revolution.

There are two dubious, or at least ambivalent, features of such promises. First, there is neglect of the possibility of addressing downstream problems directly. This is understandable because that would require change engineering and social science contributions, rather than the magic bullets that science offers, or intervention in presumed, upstream, causes. Second, and related, is the normative implication of going for upstream solutions. Consider Rachel Hurst's eloquent indictment of the new genetics in relation to the disabled (Chapter 7). Genes are seen as the source of the eventual make-up of body and mind. The possibility of fixing or eradicating impaired genes, seen as sources of impairment 'downstream', is pushed as a vision of the good life – without disablements, without disabled people. But is this suggesting that rather than trying to improve the life of the disabled, they should be eradicated?

The possibility of changing the environment rather than changing the disabled (cf. Hurst, 2006, p117) is a real alternative, but it is not discussed because it would undermine the promises. The promoters of human enhancement complain that their promises are treated selectively: bicycles are accepted, but chemical enhancement (pills) isn't (this is how John Harris phrased it in the Princeton lectures; see also Harris, 2007). In other words, the promoters treat 'enhancement' as a unitary concept, while there is actually a big difference between the bicycle, which helps in practice, and a chemical pill which changes something 'upstream' and which you must hope will have an effect (and no side-effects). For his equating parents sending their kids to a good school and giving them (or letting them have) pills to enhance their performance at school, the same critique holds. Pills are a 'point source' (like a magic bullet) for improved capacity – you don't have to do anything yourself; while to profit from a good school, work has to be done.

The difference between upstream and downstream does not only structure promises, in this case about human enhancement – there are also differences in governability. Upstream activities promising downstream effects cannot be regulated in terms of actual effects on practices, since by definition these have not yet occurred. So one is almost forced to go for precautionary approaches, or let everything go and have to remedy afterwards (cf. the well-known Collingridge dilemma; see Collingridge, 1980). Thus proposals for a moratorium, or the more modest but still precautionary approach of the Royal Society/Royal Academy of Engineering report (2004), have a structural side. Whether or not the actors are out to harass new science and technology, they have to go for precaution because the promises are about upstream solution. If there is uncertainty and concern, it must then also go for upstream governance.

Here it is helpful to introduce a further diagnosis of the situation, one which will allow me to propose specific governance by modulation. I have been using the word 'enactors' to refer to those who develop new and emerging science and technology and/or promote it. This description, inspired by Garud and Ahlstrom (1997), emphasizes that the promising option must be realized by actors enacting the required roles, and thus enacting the promise. Enactors want to realize the new science and technology, and its promises, and so see the world as a receptacle, often as a challenge or an obstacle that has to be overcome. Thus Richard Miller at the same conference could declare that 'the obstacles [to extending lifespan] are 85 per cent political, 15 per cent scientific' (see also Miller, 2002).

Necessarily so because of their position, enactors have a scenario perspective: this is what the world can be, and we must realize it. 'Enactment cycles' are followed (as Garud and Ahlstrom, 1997, phrased it) in which the positive aspects of the new option are emphasized and limitations are blamed on obstacles and resistance (Richard Miller's '85 per cent political'). Expectations will be raised, and this can lead to hype – as one now sees happening for nanotechnology, especially in the US. There are two enactor strategies in this respect: make big claims to mobilize sufficient resources to do a real job, hopefully realizing at least part of the claims; or be modest to avoid disappointment and loss of credibility. In presentations and media interviews, the two can be combined. See how Z. F. Cui, writing about regenerative medicine in this volume, concludes, 'please be patient, the future is brighter'.

The enactment cycle may not be productive, however: other actors may not follow the roles offered them in the scenario. After the experience with biotechnology in the agrofood sector, this possibility haunts nanotechnologists, and they have modified their enactor cycles to include stakeholder and public engagement. The perspective from the 'other' side is different: actors like consumer and civil society groups, but also professional agencies like the US Food and Drug Administration, are not bound to any particular new option, but can compare and select among a number of options (including selecting none at all). Their 'selection cycles' can interact with 'enactment cycles', concretely or virtually, through mutual projections. The net effect is a *de facto* governance pattern. This can be an acceptable pattern, but is not necessarily good governance of new and emerging science and technology.

Garud and Ahlstrom (1997) observed the occurrence of bridging events between the two cycles, where actors could 'probe each other's reality', without necessarily agreeing. This finding can be turned into an approach to improve governance of new and emerging science and technology. Bridging events can be organized, for example as a stakeholder workshop with enactors as well as comparative selectors (and third parties like insurance companies). 'Constructive technology assessment' (CTA) (Rip et al, 1995; Schot and Rip, 1997) has taken up this possibility, primarily to broaden the enactment cycles in terms of actors and aspects that are taken into account. To support such workshops, and in a way that speaks to the perspective of enactors, socio-technical scenarios of possible futures are developed by the organizers (which requires original research into the dynamics of the technological developments and their possible embedding in society) and discussed during and after the workshops (Rip and te Kulve, 2008).

PhD student Douglas K. R. Robinson and I have conducted a series of workshops on areas including drug delivery and molecular machines for the EU Network of Excellence Frontiers. The first, in June 2006, focused on single-cell (on a chip) analysis. These workshops are insertions in the ongoing dynamics of the Network of Excellence, and thus also in the development of the areas focused on. Substantial interaction between actors in interactive scenario workshops has challenges of its own, however. The organizers of the workshop need to reach the actors, in particular the nanotechnology actors and those allied with them, first to get them to participate at all – so there must be something at stake for the prospective participants – and second by linking up with their worlds, without completely instrumentalizing CTA. This also implies that we have to accept some of the limitations prevalent in the nano-world, in particular the concentric bias of the enactment cycle.

The challenge of the development of an integrated cell-on-a-chip platform was recognized by enactors and motivated them to participate in the workshop. Our contribution was in the preparation, including three scenarios showing possible paths (and forks in the paths), and animation during the workshop and evaluation afterwards. One effect was the recognition of the limited malleability of the cell-on-a-chip field and how difficult it may be for an enactor to make a difference. This, of course, was not new to the participants, but the scenarios articulate this 'degree of difficultness' via sketching ongoing processes and (emerging) irreversibilities. The scenarios also allowed consideration of broader aspects (for example possible reduction of animal experiments). A start-up company which participated in the workshop is now applying a version of the multi-path mapping tool. For CTA, in this case and in general, there is an uneasy trade-off between linking up with existing enactors' perspectives (so as to keep the enactors involved) and broadening their perspectives (so as to induce some change).

Such workshops provide, of course, only a limited contribution to governance. But they are productive because the agonistic interaction of the heterogeneous actors involved reflects and anticipates governance challenges, especially because socio-technical scenarios are used, which introduces substance to the discussion, in contrast to what commonly happens in open-ended focus groups and other attempts at communication and articulation. For nanotechnology, such workshops have become a recognized feature in the nano-world, at least in Europe. In other words, they are not just a tool that can be used to broaden a specific enactment cycle. Such workshops, and other interactive processes (see, for example, Marris et al, 2008), are an expected part of the process by now, and thus turn into a component of governance.

This emerging institutionalization of interaction does not seem to happen in the world of life extension and enhancement technologies – a missed opportunity of governance by modulation to improve technology and society. One can speculate why there is such an exclusive focus on promises and enactment cycles. There is one feature of the discussions which has struck me as a possible explanation (even if it is not specific to life extension and enhancement technologies). New and emerging science and technology are positioned as being introduced into society, coming in from the outside as it were (Swierstra and Rip, 2007). Thus the technology is seen as having agency by itself, while society (societal actors) must come to terms with it. Enactors can then speak in the name of the exogenous technology, and feel justified in imposing their view of progress on society – up to blaming obstacles to such progress on phobias: nano-phobia (Rip, 2006) and, in the case of life extension, gerontologiphobia (Miller, 2002, p170).

Outsiders to technology, especially when they are critical, can easily fall into the trap of attributing agency to exogenous technology. Sarah Franklin (2006, p87) shows this when she says that the view of Fukuyama and Habermas, regarding genetic manipulation as '*a force unto itself*, hostile to social order and integration', is summarized in Habermas's comment that '"biotechnology" is attributed a sinister agency'.

Thus, outsiders picturing new technology as an independent force become part of an unholy alliance with insiders, perpetuating the myth of exogenous technology.

It is important to endogenize technology, in other words to see it as part of societal dynamics. Steve Rayner, during the conference, made a similar point when he criticized the dualistic view of technology as somehow separate from society. Studies in the history and sociology of technology have shown, in abundant detail, how technology and society are a 'seamless web' (Hughes, 1986). One implication is that there is no point source from which effects and impacts emanate. All sorts of factors and interactions play a role; effects and impacts are co-produced. Our use of socio-technical scenarios in our CTA exercises are aimed at enactors in interaction with 'outsiders', but by highlighting technology's embeddedness in society, this also undermines the easy projection of agency onto the new technology.

This leads me to my concluding comment. New science and technology emerge in an existing socio-technical landscape and are shaped by it while also transforming it. Assessment and governance of new science and technology should thus not look at them in isolation, but focus on consequentialist checking of impacts, attributing utilities and then taking measures. A socio-technical landscape enables and constrains actions and interactions, and in that way functions as part of a *de facto* constitution of our society. Life extension and enhancement technologies should therefore not be addressed in terms of impacts and choices/preferences, but as part of an evolving constitution.

The basic question is what sort of world do we, and can we, live in. In particular, what sort of governance arrangements can we set up and maintain with respect to new and emerging science and technologies? There is no simple answer, but there are entrance points if one forgets about centralized governance and its illusion of control. Instead, the modulation of ongoing dynamics is possible. In this chapter, I have indicated one way to do this, and linked it to characteristic features of new and emerging science and technology.

26

Governing Our Future Selves

Daniel Sarewitz

Tomorrow's people will not arrive with a fanfare and a declaration of 'We're here!'. They have been sneaking up on us for tens of millennia and are even now carrying us along with them. To paraphrase Walt Kelly's *Pogo*, we have met the artificially enhanced and they are us, and will continue to be us, more or less. To enhance or not to enhance is a question that has decisively been answered in the affirmative.

Yet my point is certainly not that future enhancement of human capabilities is going to occur along a path predetermined by technologies unfolding according to some immutable internal logic, beyond the influence of human intentions, unresponsive to human dilemmas. Technologies are human creations, and they reflect the decisions that human beings make within organizations and cultures that are also human-made. Despite the many rather confident predictions offered in this book, the future of human enhancement is still to be created. But the technological and ethical commitment to enhancement – the transhumanist position, so strongly represented here – seems to have at least one thing right: *Homo sapiens* appears compelled as if by its very nature to try to improve its performance standards across almost every dimension of its activities and using all means at its disposal.

In this chapter I am using the term 'governance' to describe the processes by which human beings mediate their compulsion to enhance. Governance of course includes formal government policy and regulation, but it also includes a much wider array of influences and activities that can steer the directions of scientific and technological advance and uptake in response to people's preferences and choices. Governance is an outcome of social negotiations, implicit and explicit, over ends and means, over competing views of what the future might look like and should look like, and over competing understandings of what decisions are most likely to achieve what is desired. The enhancement enterprise, it turns out, is highly governed – it could not be otherwise.

Above all, perhaps, there is sport. Controversies over drug use in sports like bicycle racing and baseball are negotiations about the appropriate enhancement of human capabilities, and they are the stuff of the daily newspapers and blogs. These controversies are about cultural preferences: when athletes use certain types of drugs these days, it feels to many people like cheating, and so those drugs are forbidden. But of course many sports figures, striving for a small advantage that can be the difference between failure and fame, are still tempted, and go to great lengths to hide their illicit efforts. Pumping iron, drinking coffee, taking vitamins, using pain-killers – all these are deemed OK. Blood doping, taking human growth hormone or anabolic steroids, corking your bat, or putting a small motor in the axle of your bicycle wheel – these are not. Artificial limbs? Well, it depends. Genetic modification? Yet to be adjudicated on. The barrier separating what's right from what's wrong is fuzzy, somewhat arbitrary and likely to evolve considerably in coming years.

Fuzzy barriers are where the governance action is. Much governance activity occurs around the fuzzy barrier between therapy and enhancement, between the abnormal and the normal. Many performance-enhancing drugs can be legally obtained only with approval of a doctor – a strong form of governance. But who deserves such approval? A doctor can prescribe Ritalin to improve the performance of a child with attention deficit disorder (ADD); he or she is not supposed to prescribe it to cure my own inadequate attention span, shattered as it is from some combination of email and advancing age. The child is abnormal; I, apparently, am not. But this could change. Students, of course, seek Ritalin and related drugs to enhance their ability to study and take exams, so they have to get it illegally unless they can convince a doctor they have ADD, or find a doctor – and apparently there are many – who is willing to prescribe the drug for 'off-label' uses (see Turner, Chapter 18). These are the sorts of stories told throughout this book; they are stories of governance, of how societies make decisions about when it is OK to enhance, and when it is not, based on fuzzy and ever-shifting boundaries between categories – normal/abnormal, sick/well, ethical/unethical – that themselves are at least partly, and often strongly, culturally determined.

Societies generally seem comfortable enhancing the abilities of people whose functions are less than what is deemed normal, but are so far tending to draw the line at enhancing abilities that already fall within the normal zone. There are many exceptions, of course: for example, some women take hormone replacement therapy to avoid the effects of the normal process of menopause. And the normal zone is continually shifting; the very existence of an enhancement technology can turn what had been normal into the subnormal. When pharmaceutical companies developed the capacity to synthesize human growth hormone, shortness went from being a statistical phenomenon to a treatable condition. Unevenly spaced teeth used to be normal; braces have turned them into an aberration from the norm. Some types of enhancement are even the subject of policy mandates: in some countries, refusing to enhance your children's immune systems using vaccines is illegal.

The locations of our fuzzy boundaries evolve in response to new enhancement technologies, but the boundaries also help determine which technologies get created, at what pace and for what use. Bostrom and Sandberg (Chapter 17) and Nutt (Chapter 20) tell us that pharmaceutical companies are reluctant to pursue research and development on drugs that might improve human intelligence, memory or happiness, because they fear that government regulations will prevent the sale of such drugs. Here, in the fuzzy zone between the normal and the abnormal, is an instance where cultural, legal and economic preferences of high technology societies converge to govern the pace and direction of technological advance and human enhancement. Market opportunity is a powerful governor of enhancement. Vaccinating against infectious disease has been a powerful and effective enhancer of the human immune system (Zhao and Ma, Chapter 23), but since most remaining infectious diseases are concentrated in poor countries, where the potential to make profits is restricted, investment in discovering and developing new vaccines is restricted as well.

The science behind tomorrow's people is also being governed and steered by power struggles among scientists. Jay Olshansky and colleagues (Chapter 12) and Richard Miller (Chapter 13) insist that the obsession among mainstream biomedical scientists with understanding and curing individual disease has resulted in an unconscionable neglect of research on ageing – an alternative path of knowledge creation that would, these advocates insist, itself lead to a radical increase in disease-free life expectancy. But, as de Grey shows (Chapter 11), even within the anti-ageing domain there is lively disagreement about what counts as good science and who should get the resources to conduct it.

Societies, it turns out, have many reasons to want to govern human enhancement. Management of the health risks of new technologies is a reason that few would disagree with. Clinical trials for new pharmaceuticals, and restrictions on who can prescribe and sell them, are broadly accepted as appropriate forms of governance. But risk to some social norms also appears to be an acceptable basis for governance. We want to protect the integrity of our sports, an integrity based on some shared sense of what 'playing by the rules' should mean, and some evolving sense of how to define and enforce this. On the other hand, no one seems to care that Coleridge wrote *Kubla Khan* in the throes of opium-induced hallucination, or that, as Peter Schwartz tells us in Chapter 28, the discoverer of the polymerase chain reaction had been enhancing his imagination with LSD. Different games, different rules. And so we do govern 'recreational drugs', in part for reasons of health risk, but also apparently because society does not deem certain types of emotional or psychological enhancements to be desirable. It's OK to use antidepressants and other drugs to bring people up into a zone that experts consider to be normal, but in most societies there is so far no formal social sanction for pushing technology further to increase creativity, enlightenment or happiness. Some of the most haunting and lasting stories of techno-dystopia are those, like *Brave New World* and *Farenheit 451*, in which the citizenry are 'made' happy – happy and compliant, that is – through technology. Forced happiness somehow seems as troubling to our sense of what's right as enslavement or apocalypse.

For the most assertive advocates of human enhancement, the position on such matters seems to be that all performance boundaries (normal/abnormal, therapeutic/ enhancing, even human/non-human) are arbitrary constructs that evolve over time. From this perspective – that of the transhumanists – such boundaries are therefore invalid bases for making decisions about human enhancement, or for opposing enhancements or trying to slow them down or direct them in certain ways. James Hughes (Chapter 6) tells us that 'human nature' cannot be precisely defined or located, and therefore is an illusion and an illegitimate basis for opposing enhancement. Bostrom and Sandberg (Chapter 17) show us that, as human beings, we have always been engaged in enhancement, and thus (apparently) that claims against enhancement are illegitimate. For every new enhancement, one can pretty easily think of a past analogy that now seems perfectly acceptable. Why should drugs to increase intelligence be viewed any differently from, say, books or school? Why, that is, all the fuss over means if we agree on the ends? Caffeine, exercise and vaccination are all powerful human enhancers, and who's complaining about those? Let's just get on with the business of making ourselves better as fast as possible.

I find this an irritating position, for three reasons. First, organized society is simply impossible without the fuzzy boundaries that we use to make sense of the world and build systems of rules that ward off chaos. The fact that the boundaries cannot be anchored in natural law, and evolve with time, does not speak at all to their value for structuring society. Even the most universally accepted boundaries, such as proscriptions on killing, have their exceptions and their cultural and historical contingencies. Second, the current state of human enhancement is itself the result of a long, complex history of governance in which cultural preferences, economic factors and political conflict (including, perhaps above all, armed military conflict) have interacted to shape how technologies evolve and are used today. The fuzzy boundaries that help to govern our thinking about and use of technology evolve in response to these dialectics, just as interpretations of laws evolve over time in response to changing social conditions. And third, therefore, the future of human enhancement is not going to unfold in some organic or optimal way, but will reflect and respond to exactly the sorts of concerns, dilemmas and disputes aired in this volume.

The key governance problem, it seems to me, is not to do with the negotiation of boundaries per se, which is just what people do, but concerns the social conditions and assumptions that strongly define both the terms of the discussion and the trajectories that future enhancement will actually follow. Yet, as Nordmann (Chapter 3) points out, the scenarios of future enhancement that seem to inspire the transhumanists – many examples of which can be found in this book – are offered with little if any reference back to today's world. In this sense they belong much more to the category of science and technology boosterism than reflections on the meanings of our present world for the future. One might reasonably wish for the richer imaginings of good science fiction writers, who extrapolate not disembodied technological trends but technologies continually conditioned – governed – by the ongoing dilemmas and uncertainties of the human condition.

What about, for example, justice, equity and power? Decisions about, access to and benefits from new technologies tend to concentrate among particular segments of society with economic and political power. Responsible discussion of the future prospects and promises of human enhancement must consider who stands to win and who stands to lose from the particular decisions that are made about how and who to enhance – decisions being made now. Enhancement advocates like Savulescu (Chapter 22) and Hughes (Chapter 6) argue that the nature of enhancement itself seems to militate in some necessary way towards justice and equality, but Hurst (Chapter 7) and Wolbring (Chapter 24) provide arguments and instances of the contrary, as does the worldwide distribution of cosmetic surgery, nanotechnology-enhanced exoskeletons, organ transplants and so on. One might also consider the distribution of dis-enhancements like obesity, drug abuse and illicit organ harvesting, also concentrated in particular ways. Or the fact that the locus of the most aggressive enhancement research is probably the military, as Garreau discusses in Chapter 2. Wouldn't it be great if pilots could control their jets with their fast thoughts rather than their slow hands? From the military perspective, the goal is precisely to create asymmetries in power; the trajectories of transhumanism cannot but inherit these asymmetries. Visvanathan (Chapter 5) notes that:

> *The debates on life enhancement and longevity belong to … discourses on progress, growth, development and perfection. They look linear and morally innocent unless we confront their obverse side in the world of triage, obsolescence, waste, the defeated and the broken.*

These sorts of initial conditions are the agar-agar from which tomorrow's people will sprout.

Yet our discussions about tomorrow's people seem often to involve a leap to a future condition that derives solely from claims made on the basis of emerging technological specifications and capacities, a sort of virgin birth unsullied by the conditions of the world today. Nor is governance anywhere to be found in the discussion of these technologies, unless it is located in people's irrational resistance to change, as Silver argues in Chapter 8. But governance is not a choice – it is a condition. Tomorrow's people will emerge not just from the consequences of research and development, but also from the influence of customs, norms, laws, argumentation, economics, politics, science fiction and perhaps even philosophy. This is not to say that the path from here to tomorrow's people can be accurately surveyed and laid out in advance. But the decisions about what research to do, and what to do with that research, are right now being negotiated in discussions like the one taking place in this very book.

Perhaps this portrayal of governance seems too modest, too unconscious, too passive? Is there no way to do better, to steer more decisively in directions that will benefit society, and steer away from threatening or dangerous or divisive paths? One obvious problem, of course, is that people may – and usually do – disagree about what might actually be a benefit or a threat. And even if they did agree on these ideas, they are likely then to disagree about how to chart a course in the desired direction. Better governance is not about controlling where the technology goes by prior agreement (an impossibility), but about bringing into the open the decisions that are being made about human enhancement, in venues from the laboratory to the corporate boardroom to the houses of government, and subjecting them to serious, open and pluralistic reflection. But better governance also depends on the attributes of the societies into which human enhancement technologies are introduced. The consequences of the distinctively human struggle for a more just and equitable society among today's people is likely to be the strongest determinant of the type of world inherited and inhabited by tomorrow's.

27

Global Population Ageing and the World's Future Human Capital

Wolfgang Lutz

A DEMOGRAPHICALLY DIVIDED WORLD

Current global demographic trends and the associated challenges are somewhat confusing to many observers. On the one hand, in many developing countries population growth rates are still very high due to birth rates well above replacement level (of two surviving children per woman) and a very young age structure. For this reason, in a number of countries the population is likely to double over the coming decades. On the global level, this leads us to expect that the world population will increase from its current 6.4 billion to somewhat below 9 billion by the middle of the century. On the other hand, there are an increasing number of countries in which the birth rates have fallen well below replacement level and the population is ageing rapidly. For these countries we expect a future of even more rapid population ageing and, in many cases, a shrinking of total population size. Because of these significantly different demographic trends, for some parts of the world there is still concern about the negative consequences of rapid population growth, while simultaneously in other parts there is concern about the negative implications of rapid population ageing.

DEMOGRAPHIC TRANSITION AS THE MAIN DRIVER

Explanations and projections of fertility trends in different parts of the world have been generally guided by the paradigm of demographic transition, which assumes that after an initial decline in death rates, birth rates also start to fall after a certain lag. In this general form, the model has received overwhelming empirical support in capturing the remarkable fertility changes that happened over the 20th century.

The demographic transition began in today's more developed countries (MDCs) in the late 18th and 19th centuries, and spread to today's less developed countries (LDCs) in the latter half of the 20th century (Notestein, 1945; Davis, 1954 and 1991; Coale, 1973). The conventional 'theory' of demographic transition predicts that, as living standards rise and health conditions improve, mortality rates first decline and then, somewhat later, fertility rates decline. This demographic transition theory has evolved as a generalization of the typical sequence of events in what are now MDCs, where mortality rates first declined comparatively gradually, beginning in the late 1700s, and then more rapidly in the late 1800s, and where, after a varying lag of up to 100 years, fertility rates declined as well. Different societies experienced transition in different ways, and today various regions of the world are following distinctive paths (Tabah, 1989). Nonetheless, the broad result was, and is, a gradual transition from a small, slowly growing population with high mortality and high fertility to a large, slowly growing or even slowly shrinking population with low mortality and low fertility rates. During the transition itself, population growth accelerates because the decline in death rates precedes the decline in birth rates.

However, the demographic transition paradigm that has been useful for explaining global demographic trends during the 20th century, and that has strong predictive power when it comes to projecting future trends in countries that still have high fertility, has nothing to say about the future of fertility in Europe. The recently popular notion of a 'second demographic transition' is a useful way to describe a bundle of behavioural and normative changes that happened recently in Europe, but it has no predictive power. The social sciences have not yet come up with a useful theory to predict the future fertility level of post-demographic transition societies. All that forecasters can do is to try to define a likely range of uncertainty.

MAPPING THE UNCERTAINTY RANGE OF DEMOGRAPHIC TRENDS IN THE 21ST CENTURY

All three components of demographic change – fertility, mortality and migration – are uncertain in their future trends. The UN medium variant (United Nations, 2003) is based on an attempt to make the assumptions that are most likely from today's perspective. But we already know today that there is a high probability that the actual future trends will either be above or below the medium assumption. How should we deal with this significant uncertainty in population forecasting? This question is the title of a recent special issue of the *International Statistical Review*

(Lutz and Goldstein, 2004). This state-of-the-art report shows that in the field of population forecasting, we are currently seeing a paradigm change from scenarios to probabilistic forecasting. Scenarios are descriptions of possible future paths without any statement of the likelihood of these paths. Particularly in the case of deep uncertainty, in other words when there is not only parameter uncertainty but the entire model is uncertain, they have become a standard tool for thinking about the future. Since we use a familiar and accepted model for population projection – the cohort component model – with three components of change (fertility, mortality and migration), only the parameters are uncertain. For this reason, some decades ago forecasting agencies around the world, following the example of the United Nations Population Division, produced high and low variants in addition to the medium variant. This high–low range is supposed to indicate a 'plausible range' of future population trends. But such a high–low range can only be defined in terms of one of the three components of change and is mostly based on alternative fertility assumptions, while uncertainty in mortality and migration is disregarded.

To remedy such shortcomings, the International Institute for Applied Systems Analysis (IIASA) produced the first fully probabilistic projections of the world population (Lutz et al, 1997). These were essentially based on subjective probability distributions for future fertility, mortality and migration, as defined by a group of experts. In 2001 new probabilistic projections were performed that are based on a synthesis of three alternative approaches (time series analysis, ex post error analysis and argument-based expert views) (Lutz et al, 2001). The examples below are taken from this 2001 forecast. Such probabilistic projections go in several important dimensions beyond the traditional scenario analysis: they are able to simultaneously consider the uncertainty in all three components of change; they can define in more precise quantitative terms what uncertainty intervals the given ranges cover; and based on the assumption of certain correlations, they can aggregate from the regional to the global level in a probabilistically consistent way. These important advantages of a probabilistic approach over a scenario approach make it worthwhile to consider whether in other fields, such as environmental change or future health, one should go beyond scenarios.

The key findings of Lutz et al (2001) are that, with a high probability of above 80 per cent, world population will peak over the course of this century and then start to decline and that the 21st century will bring significant population ageing in all parts of the world. In short, it is concluded that, while the 20th century was the century of population growth, with the world growing from 1.6 to 6.1 billion people, the 21st century will be the century of population ageing, with the global proportion above age 60 increasing from currently 10 per cent to between 24 and 44 per cent (80 per cent uncertainty interval). Even more significantly, the proportion of the world population that is above age 80 will increase from currently 1 per cent to between 4 and 20 per cent, depending largely on the future course of life expectancy.

Figure 27.1 shows that for Western Europe the proportion above age 80 might even increase much more dramatically than at the global level. The figure shows that currently around 3 per cent of the population is above age 80, and that this proportion will not change much over the coming decade. After 2030, however, the uncertainty range opens up very quickly. In 2050, the 95 per cent interval already goes from around 4 per cent at the low end to more than 20 per cent at the high end, with the median at around 10 per cent. In other words, in 2050 the proportion above age 80 is likely to be three times as high as today, but it could even be six times as high. Its actual level will depend mostly on future old-age mortality – whether life expectancy will level off towards a maximum or whether it will continue to increase unabated. This difference becomes even more significant during the second half of the century. By the end of the century, the 95 per cent interval is extremely wide, ranging from essentially the current level of 3 per cent to an incredible 43 per cent of the population above age 80. Even the median shows a proportion of about 20 per cent. Societies with such significant proportions of the population above age 80 will clearly be very different from those of today. Most likely, however, is that an average 80-year-old person during the second half of the century will be in much better physical health than an average 80-year-old person today.

Figure 27.1 shows two lines (at right of the chart) indicating the 'high' and 'low' variants of the 1998 UN long-range projections of the proportion of the population above age 80 in 2100 (United Nations, 1999). Since the UN does not

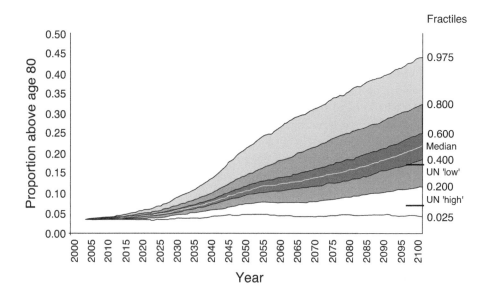

Figure 27.1 *Proportion of population above age 80 in Western Europe (UN 'low' scenario for 2100 = 0.17, UN 'high' scenario = 0.07)*

use alternative mortality assumptions in their variants, it is not surprising that the range is extremely narrow. Also, the UN seems to expect much lower improvements in life expectancy in Western Europe. This illustrates that the traditional variants approach that only varies the fertility assumptions is a highly problematic way of dealing with uncertainty, and certainly should not be interpreted, as it often is, as giving a 'plausible range'. Clearly, the future course of old-age mortality and disability provides us with many difficult but highly important research questions.

FUTURE HUMAN CAPITAL

The above-described demographic trends are currently resulting in a major shift of the distribution of the world population over different continents and regions. Europe is not only likely to age rapidly and shrink in absolute numbers, but is even more rapidly seeing its share of the world population decline. As significant as these changes in relative population size are, it is not clear exactly what they will imply for a region's geopolitical standing. The strength and influence of a nation or a continent is not directly a function of its population size. If it were, then Africa today should have a similar standing in international politics, economics or military strength to that of Europe, and this is not remotely the case. What seems to count more than the sheer number of people is the human capital, which can be defined in a simplified way by looking at the people of working age stratified by their level of education. The global distribution of human capital is changing as well, but the pattern looks rather different from that of mere population numbers.

The first global projections of human capital have been recently produced by IIASA (Lutz et al 2004). Table 27.1 lists the persons of working age that had at least some secondary or tertiary education in 2000 and gives two alternative scenarios to 2030. The 'constant' scenario assumes that current school enrolment rates stay unchanged over time, which will result in significant improvements of human capital in many countries because of past improvements in education and the process by which the less educated older cohorts will be replaced by better educated younger cohorts. The other scenario, labelled 'ICPD', assumes that the ambitious education goals defined at the International Conference on Population and Development (ICPD) in Cairo in 1994 will be achieved. These include a closing of the gender gap in education and universal primary education.

Table 27.1 shows that in terms of human capital, Europe (including Russia) is still a world power, with well over 350 million working-age people with higher education – many more than in Africa, and even more than the huge South Asian subcontinent. This helps to put the pure population numbers into perspective. But the table also shows that significant changes in the global distribution of human capital are to be expected, even under the constant enrolment scenario. Under this scenario, every world region will see some improvement of its overall human capital. On a relative scale, gains in today's least developed regions will be strongest, partly because the recent improvements in educating the younger generation have already been a significant gain, given the virtual absence of education for the older cohorts. In absolute terms, even under this constant enrolment scenario, huge gains in the number of working-age people with secondary or tertiary education are to be expected in Latin America, South Asia and the China region. In today's industrialized countries, only moderate gains are to be expected. Comparing these results to those of the most optimistic scenario, which assumes the success of the education goals of the ICPD, there is surprisingly little difference. This is due to the great momentum of educational improvement. Increases in school enrolment today and over the coming decade will only very slowly affect the average educational attainment of the whole working-age population. The difference is worth noting in sub-Saharan Africa, because the current school enrolment rates there are still far below the Cairo targets. But because the ICPD also implies lower fertility in some regions, the absolute numbers are even smaller under the ICPD than under the constant rate scenario.

Figure 27.2 summarizes the information of Table 27.1 in graphical form. It compares four 'mega-regions'. It shows that currently Europe and North America together still dominate the world in terms of human capital, although South Asia

Table 27.1 *Population (in millions) aged 20–65 by education and sex in 2000 and in 2030 according to the 'constant' and 'ICPD' scenarios*

Regions	Base year		Secondary and Tertiary Constant		ICPD	
	2000 Male	2000 Female	2030 Male	2030 Female	2030 Male	2030 Female
North Africa	19	11	47	38	49	41
Sub-Saharan Africa	32	17	79	61	106	90
North America	88	89	100	99	100	99
Latin America	66	65	140	143	143	147
Central Asia	13	13	25	25	25	25
Middle East	17	12	50	40	53	46
South Asia	134	57	250	116	288	195
China region	238	153	416	354	406	346
Pacific Asia	53	41	99	90	106	99
Pacific OECD[a]	40	40	40	39	39	40
Western Europe	106	95	124	122	125	122
Eastern Europe	26	23	31	30	31	31
FSU Europe[b]	54	57	58	61	59	62
World	887	673	1459	1219	1531	1343

Note: [a]Organisation for Economic Co-operation and Development members in the Pacific region; [b]European part of the former Soviet Union.
Source: Lutz et al (2004), p149

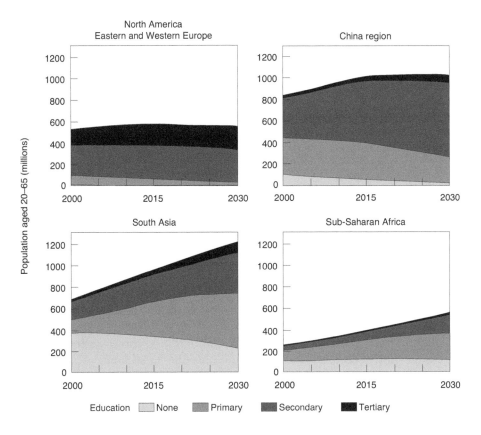

Figure 27.2 *Population (in millions) aged 20–65 by level of education, according to the 'ICPD' scenario in four mega-regions, 2000–2030*

Source: Lutz et al (2004), p138

and the China region are already bigger in terms of working-age population. The figure also shows the different pathways of China and South Asia (India), which reflects the fact that unlike South Asia, China has invested heavily over the past decades in primary and secondary education, and will see a peaking of its population size over the coming decades. South Asia will soon surpass the China region in terms of population size, but will fall back in terms of human capital. Even under the most optimistic scenario, Africa will see only very moderate increases in human capital. An interesting point worth noting is that China's human capital is increasing so rapidly, that by around 2015 China will have more people of working age with secondary or tertiary education than Europe and North America together. These global shifts in human capital are likely to result in changing geopolitical and economic weights and also have significant implications for global health and wellbeing.

Human Capital in Sustainable Development

This chapter has presented the case for adding a potentially very important new aspect to the analysis of global population trends, namely the global change in the composition of the population by education. This helps to shift the focus from exclusively studying the quantity of people to adding the quality dimension. In this context the notion of human capital can best be defined by:

$$\text{Human Capital} = \text{Population} \times \text{Health} \times \text{Education}$$

To conclude, this chapter has explicitly and quantitatively combined the population and educational attainment dimensions. The health dimension was less explicitly treated, because the ideal data – age- and sex-specific disability rates – are not available for the majority of the world population. But through the consideration of life expectancy, health is implicitly in the calculations and, after all, educational attainment and health tend to correlate strongly. An example of how disability and health can be explicitly incorporated into this multi-state population projection framework is given in Lutz (2006).

In addition, a better-educated population has been shown to have greater economic growth potential and to be less vulnerable to all kinds of new health threats resulting from environmental change or other global influences, as well as showing a greater adaptive capacity. This whole body of literature cannot be discussed here, but deriving from it, there is reason to believe that human capital can indeed be considered to be the ultimate resource for sustainable development. While we are in danger of running out of many resources, ranging from natural to financial ones, and while we in Europe may even run out of young people, we can always learn more. There seems to be no limit to improving our skills and human capital. This factor may become even more decisive as our technologies become more complex and our lives last longer. The bad news is, however, that some countries are clearly falling behind in the improvement of their human capital and thus are compromising their chances for the future.

Part X

Postscript

28

Choosing our Biological Future

Peter Schwartz

The final chapter in this volume offers a look back from mid-century at how we mastered our own evolution. My contribution is dedicated to Mary Douglas, who we now know was one of the most far-sighted people of the 20th century, with special thanks to James Martin and Bill Sharpe, who reached out in time and managed to capture me.

I address you in this chapter from my vantage point as a 104 year old here in the world of 2050. As I look around my world, I see that we are in a period of accelerating change, with a new biological era driving development. The jump to the biological age in which I live is as great as the leap from the agricultural societies to the industrial and information age, from the time when we used muscles and wind and water as the principal sources of energy to the remarkable industrial and information age which all of you live in to the biological age in which you all live. Indeed, the change is even greater than that because, of course, in that era the change was mostly about the outside world and, as Joel Garreau observed, it became about changing the inside world.

And it's important that you try to imagine a much wealthier world, which fortunately, as Robin Hanson suggested we would, we currently enjoy. Unfortunately, Lord Richard Layard was wrong when he said that economic growth was no longer needed for happiness. That might have been true for the well-off, but the opposite was even more true – the absence of growth led to misery. Fortunately, the middle class of today have a per capita income roughly ten times the income of the middle class of your time. There are still poor in the world, though – about a billion in those parts of the world disconnected from the advances of modern science.

Some have been slow to develop, like Africa and the Middle East, which is now poor as the oil is beginning to go, and others have opted out for reasons of their religious beliefs, leaving large pockets of poverty in the world.

But returning to bioscience. The impacts have varied enormously around the world, as the constraints on technology have proved to be very different. The US, China, the UK and India all went in different directions. As James Martin (2007) has suggested, we needed to reinvent society, but there turned out to be many reinventions, not just one. Some societies chose to constrain the most extreme capabilities, so changed very little. They dealt mostly with disease and the maladies of ageing. There were no novel enhancements, no life extension, and they stopped along the way. They chose that exit. Indeed, different societies chose different exit paths along the way. So today, when we look around the world, we can see in fact different moments in time, all existing together. But others chose to go much further, either by choice or by the inability to constrain the choices in technology.

Indeed, it came to be acceptable to modify as the very visible plastic surgery of ageing actors and rock stars moved on into enhancements, becoming cool and fashionable. So we see very youthful old people today. Centenarians, like me, could fool you into thinking we're 60. And real ageing reduction is already a reality, so our 60 year olds, like my son Ben, would probably look about 30 to you. And we're seeing the beginning of real life extension. The first people, though still only a very few, are beginning to live beyond 120; however, I expect to make a very robust 150, so I see at least another 50 years in front of me. And we're beginning to see, of course, bio-improvements of all sorts.

Most people have opted for improved memory, sight and hearing. And almost all physically demanding tasks today have technological leverage, either in the form of performance enhancing drugs or external body modifications. And we're even beginning to see some new human species. Not many, but a few. The new capabilities of our new people – smarter, stronger, bigger, faster, more advanced in improved senses, for example with eagle eyes or infrared or UV eyes, disease resistant and longer lived – are being locked in at the genetic level.

For many, our enhanced capacity for learning and for deep reflection that cognitive enhancements have enabled us lead us to many more eureka moments, as Susan Greenfield (2003) has urged. Or, for many more, moments of transcendence, as David Nutt suggested was the definition of happiness. I'm fortunate – I have my Mary Douglas 3.0 chip installed. We managed to capture her consciousness, and her wisdom asks me very hard questions all the time and keeps me on the right path.

On the whole, the world is full of happier, healthier, longer-lived, richer people living in a more sustainable ecosystem in a highly productive economy. People living not merely extended lives but expanded lives, as my wife Kathleen has put it. They experience the world and their relationships, as Nick Baylis argued, more deeply and more broadly. So life is not only increasingly long but increasingly broad and deep as well.

But I have to admit that my world is no fairy tale. There are clouds on the horizon. A few places in the world have experienced the Bill Joy 'Hellworld' that Joel Garreau discussed, with its horrific consequences. These are the bio-era equivalents of Chernobyl and Bhopal. And we are seeing the growing potential for conflict between the old and the new humans and, indeed, in the eyes of some, the new humans are an abomination in the eyes of God. Passions are aroused by these new realities. Is this how evolution happens, as we take hold of it? Will the new humans gradually wipe out old ones like me? Maybe this is how it is supposed to be.

Well, how did this remarkable transformation come about? Let's take a look back at some of the socio-technical developments of the first parts of the 21st century. To construct these scenarios, as Arie Rip suggested, we really need to begin with the science. And indeed, we see that science was driven forward by three things: by new tools, by the sources of funding and by the scientists themselves.

And at the end of the 20th and beginning of the 21st century we saw huge leaps in new tools. Among the most famous, of course, was the polymerase chain reaction (PCR), which made possible the new era of genetic modification. This was itself the product of one of the first enhanced discoveries. As the well-known story goes, Kerry Mullis, the discoverer of the PCR, found it one day (or discovered it and imagined it one day) while driving in his car on his way to Mendocino, up Highway 128, winding along the curves of the coast. And suddenly he saw in his mind's eye that helix beginning to unzip and rezip, and he figured out how to in fact multiply DNA. Of course, as it happens, he was stoned on LSD, completely out of his mind, and thus the vision emerged out of an enhanced mind. Now, unlike the athletes of the time who have little asterisks next to their records, there is no asterisk next to his Nobel Prize.

There were genomic information systems, new modes of sensing, such as functional MRI and, of course, in our time, the advances in quantum computing. And, as we know, as in architecture so in science, form follows funding. So we had the war on cancer, we had the human genome project, pharmacogenomics, in California the stem cell initiative, in Singapore the new National Research Foundation. Where governments and society chose to put their money, scientists followed.

But for scientists, the real tipping point came in the 1990s, as science came back into fashion and huge rewards became available for successful sciences in the new business start-ups. And so we saw a huge increase in the number of scientists, the quality of scientists and the locations all over the world, no longer just in the West. And, of course, different disciplines were in favour. I, unfortunately, was a product of the Sputnik era and got my degree in astronautical engineering – I was once a rocket scientist – but now it is, of course, biology that is in favour. And it was the interests of these biological scientists with their ruthless curiosity and their dreams of a better world that drove science forward.

So how far did the science go? Well, it could have been a world of minimal progress. It could have been, as in fact many feared, that the scientific challenge was like AI or fusion power or the space programme, where we spent lots of money but we made relatively little progress. Not because we're stupid, but because they were very hard problems. And it turned out that putting the sun in a bottle proved to be very hard – it is still 40 years off, just as it was in 1960.

There were many dead ends for a while. Stem cells proved harder to tame. Gene therapies proved more difficult to tame. So there was very little gain at first in our understanding of the brain. In fact, it led to a biotech bust – there was reduced public and private funding. Things could have stopped there and there may have been no biological era. That would have been one scenario, a scenario in which history would have been very different and our world would not have changed very much from yours, trapped by the failures of our ability to carry science forward. But fortunately it didn't end there. In fact, we began to make some modest gains, if slowly at first. And it was a bit like the prevail scenario that Joel Garreau described. At first, it was mostly about traditional strategies improving: some stem cell success, some tissue engineering, a little bit of genomics, and, as Tom Kirkwood suggested, ageing was changing but slowly. However, as Aubrey de Grey suggested, we were beginning to postpone frailty, and, as Jay Olshansky suggested, we were beginning to reap the longevity dividend.

But it was mostly about better disease therapies for more diseases and reducing the effects of ageing, improving the quality of life in old age, living longer younger, better sight, memory, skin, bone, muscle strength, joints. But there was almost no improvement or enhancement, except for in the military, the security forces and, of course, the criminal sector and for the handful of athletes who chose to break the rules.

We could have stopped there, and that would have been another scenario, but fortunately we didn't. In fact, we moved on into the new biological era, a bit like the world described by Ray Kurzweil (2005). Because it was an era of high potential rewards, drawing talent and investment, continuing successes fed on each other, amplifying the rate of change.

Indeed, we saw a major intellectual leap in the first decades of the 21st century, from 2006 to 2030. This resembled the world of physics from 1905 to 1930, when the discoveries of relativity, quantum theory and most of modern physics had emerged. Indeed, as the great scientist William Bainbridge suggested, it was an era of convergent sciences, as physics, biology, chemistry and mathematics converged at the very small scale. And new discoveries came from surprising places. Now, fortunately, most of us who had iPods now use the iHear to compensate for our loss of hearing from all those iPods.

What we saw was increasing capacity to intervene and control to become ever more effective. Indeed, we developed a new mathematics of biology, as mathematics had grown as the handmaiden of physics in the 18th, 19th and 20th century mathematics became one of the tools of biology in the 21st century as we learned how to deal with cell epidemiology and morphology. We gained the ability to control the trajectory in the development and growth of cells. We were able to intervene at the molecular level, the cellular level and the organismic level and reliably genetically modify current organisms and entrain genetic change. We've gained the ability to cure or minimize almost all diseases and repair and replace damaged and deteriorating tissues and organs. And most of all, we've been able to maintain high-level system functioning.

We have what you would call super-health: a high level of vigour, not easily fatigued, need less sleep, a great capacity for pleasure and joy, even the ability to improve vision, memory, cognition, strength and speed. Some have new human capabilities: a few of us have learned how to breathe underwater and the first spaceships are on their way to colonize Mars, with the first people modified to breathe Martian atmosphere. Indeed, the plans are just being laid to download the first minds into the first sublight starships. Undoubtedly Bill Bainbridge and I will have our virtual clones onboard, on our way to the stars, fulfilling our dreams.

And what we gained along the way in that scientific scenario was the ability to extend human lifespan. First marginally, by a decade or two, and then even longer, and we are pretty confident that we see no end in sight. So by 2030, humans were already experiencing a great deal of change and they began to redesign the humans of the future. First, improved humans were coming along. Genetically and otherwise modified human beings are now being born. We see ageing reduction therapies being practised by the young well-off with the goal of radically reducing the rate of ageing and extending life.

So how did society change in the face of all of these advances? Well, in the early days of minimal progress, there was minimal impact. The issues were mainly about access to different therapies and perhaps one or two important ethical questions on topics such as stem cells, but the big questions and the big changes existed only in the imagination of your time. It could have been a big bust, but it wasn't. And as our technologies and capabilities increased, the impacts became more dramatic.

At first, the largest impacts were in the youthfulness of the elderly, extending productive life and reducing the costs of ageing. And even at the turn of the century, change was already showing up in the baby boom. Healthier, longer, as the demographic projections of Wolfgang Lutz suggested. As a result, we were gaining improved social productivity from healthier people and reduced disability. But there was some spin-off towards improvement from these walled-off and illicit technologies, mostly among the young, strange variants. You thought piercing and spiky hair was something; that's nothing to the implants that came with it.

There were internal tensions over the use of these early technologies. Should a child take memory enhancers before their school exams? Should your surgeon take them? The value of controversial therapies like stem cells created a great deal of social conflict, but the prize was worth fighting for. And these were the steps along the way to the new biological era that I live in today with my centenarian wife who doesn't look a day over 50. So how did you decide? How did you make these choices to produce the world that we experience today?

Well, you thought about it in different ways, using different frameworks and applying different philosophies for observing and deciding. Each led to different sets of significant observables and different conclusions of desirability. These answered the different questions of what was important, allowable or compulsory, and what must be prevented. Some took a highly individualistic approach. They were libertarian, market-oriented – as long as I do no harm, I should be free to do whatever I want. Others were more egalitarian – it's good if we can all have it and bad if we can't. For many, it was a religious matter – good if my faith, God, book, tablet or scroll approves and bad if they don't.

Some have quoted that ancient wise man, Jesus of Galilee, when he said 'what shall it profit us if we gain the world and lose our soul?'. But they have little to say to those of us who didn't share their joy in the suffering of this ancient teacher while being nailed to the cross. The economist said it was good if it increases economic growth and productivity and bad if it slows growth or reduces productivity. The ecologist said it was good if it leads to increased sustainability and bad if it decreases ecosystem robustness.

But each framework assumed its own primacy and tolerated the others in varying degrees. The question of whether your God likes my new super-smart brain doesn't matter to a libertarian – if I don't harm you, no problem. But a fundamentalist Christian might say, 'Your existence offends my God; you are an abomination in His eyes!'. Egalitarians might say it's OK if we can all get the improvements and not if only I and the other privileged few can have them. An economist would have suggested that the prices will eventually fall and access will improve so the early movers are helping the latecomers.

Fortunately, we had a great opportunity to learn, because there were in fact many great natural experiments going on around the world, as many different nations adopted different frameworks for addressing these issues, and knowledge and capability of course quickly spread round the world in the Wikipedia era.

Shiv Visvanathan talked about multiple myths, and indeed we've had multiple myths – more, as he suggested, than Prometheus and Faust. We've had Kali, Shiva, Quetzalcoatl and, for many, the Jesus, Moses and Mohammed myths – all the myths had their adherents. And, as a result, we saw different paths in the US, the UK, France, Sweden, Italy, Hungary, Cuba, Russia, India, China, Japan, North Korea, South Korea, Singapore, all over the world. Each now has different capabilities as a result of the different societal choices they made.

So we had the opportunity to learn from their successes and failures. But which societies managed their choices well on behalf of their people and which failed? Well, failure meant different things. For some, it was failing to capture the benefits of the new capabilities. For others, failure was embodied in abusing the new capabilities, access by only a corrupt elite, or the indirect unanticipated consequences of the technology in the form of social divisions or new biological problems (the equivalent of thalidomide or DES (diethylstilbestrol) or ecosystem effects). For others, success has also meant different things. In some cases, it was about the benefits for the individual; for some, it was about the benefits for society; and for others it was a question of preserving what was believed to be sacred.

Now, in hindsight, we can see how people saw the balance of risks. They saw far more gain than pain. Indeed, the comments on real happiness that were made at the time were mostly by well-off educated people with continuing good prospects for themselves and their children. They didn't ask those who were suffering and dying or worse. But even more people wanted to live well and long and didn't mind using all the tools of self improvement to be a bit taller or a bit slimmer or see a bit better. The majority saw more upside potential than downside fears. However, the science proved very hard and moved far more slowly than many of us would have liked, and many philosophers and social critics got ahead of themselves in many of their ethical discussions. For example, in the debate surrounding stem cells, they were overly squeamish over only imaginary ills.

In the end, we fulfilled our ethical obligation to find out what was possible. We were not deterred by those who feared as yet only imaginary impacts. We pushed back the frontiers of knowledge in the spirit of Freeman Dyson (1979) and his 'infinite gain'. And as Kevin Kelly, the brilliant observer of science, author of *Out of Control* and, I might add, a fundamentalist Christian, said in his off-quoted lecture from long ago:

> *Science is the way we surprise God. That's what we're here for. Our moral obligation is to generate possibilities, to discover the infinite ways, however complex and high dimension, to play the infinite game. It will take all possible species of intelligence in order for the universe to understand itself. Science in this way is holy. It is a divine trip.* (Kelly, 2006)

So if it was biologically possible, then it happened. It was unstoppable but it just took a very long time. Changing human biology is happening. We are now inevitably the masters of evolution. For the people of our time, we face the need to be wise about our evolution. Indeed, wisdom is now a survival trait.

In conclusion, fortunately for us, the people of your time fulfilled your sacred obligation to maximize the pace of improving human capabilities. But you had very clear priorities. You began with reducing suffering, for which there seemed to be no downside. Then you gradually extended the quality of life for the elderly, and again there appeared to be no downside. But, before too long, you began to extend human lifespan. Now, I don't think there was any downside, but for some there was, and there was a reaction against it. And then you began to improve human performance. For some this was a major question but, fortunately, we answered these questions wisely and as a result we live better.

So now let me return to my time, and you to yours, having had a preview of how your lives' choices led to the people of your tomorrow and my today. Because those choices were wise, the world is a better place and, most important of all, our children will be playing the infinite game as far off into the future as we can imagine.

References

Academy of Medical Sciences (2008) 'Brain science, addiction and drugs', report of a working group chaired by Professor Sir Gabriel Horn, FRS, FRCP, Academy of Medical Sciences

Alteheld, N., Roessler, G., Vobig, M. and Walter R. (2004) 'The retina implant: A new approach to a visual prosthesis', *Biomedizinische Technik*, vol 49, no 4, pp99–103

Alzheimer's Disease International, 'Alzheimer's disease annual report, 2004–2005', www.frost.com/prod/servlet/dsd-fact-file.pag?docid=38565311, accessed 20 May 2008

Anand, S. and Ravallion, M. (1993) 'Human development in poor countries: On the role of private incomes and public services', *Journal of Economic Perspectives*, vol 7, no 1, pp133–150

Anderson, C. et al (2003) 'Manifesto on biotechnology and human dignity', www.cbc-network.org/redesigned/manifesto.php, accessed 20 May 2008

Annas, G. (2001) 'Genism, racism, and the prospect of genetic genocide', presented at the World Conference Against Racism, www.thehumanfuture.org/commentaries/genetic_discrimination/geneticdiscrimination_commentary_annas01.html, accessed 20 May 2008

Annas, G. (2002) 'Cell division', *Boston Globe*, available from http://genetics-and-society.org/resources/items/20020421_globe_annas.html, accessed 20 May 2008

Annas, G., Andrews, L. and Isasi, R. (2002) 'Protecting the endangered human: Toward an international treaty prohibiting cloning and inheritable alterations', *American Journal of Law and Medicine*, vol 28, nos 2–3, pp151–178

Antal, A., Nitsche, M. A., Kincses, T. Z., Kruse, W., Hoffmann, K. P. and Paulus, W. (2004a) 'Facilitation of visuo-motor learning by transcranial direct current stimulation of the motor and extrastriate visual areas in humans', *European Journal of Neuroscience*, vol 19, no 10, pp2888–2892

Antal, A., Nitsche, M. A., Kruse, W., Kincses, T. Z., Hoffmann, K. P. and Paulus, W. (2004b) 'Direct current stimulation over V5 enhances visuomotor coordination by improving motion perception in humans', *Journal of Cognitive Neuroscience*, vol 16, no 4, pp521–527

Arendt, Hannah (1958) *The Human Condition*, University of Chicago Press, Chicago, IL

Arking, D. E. et al (2005) 'Association between a functional variant of the *KLOTHO* gene and high-density lipoprotein cholesterol, blood pressure, stroke, and longevity', *Circulation Research*, vol 96, p412–418

Arnhart, L. (2003) 'Human nature is here to stay', *The New Atlantis*, no 2, summer, pp65–78

Arnsten, A. F. T. and Robbins, T. W. (2002) 'Neurochemical modulation of prefrontal cortical function in humans and animals', in D. T. Stuss and R. T. Knight (eds) *Principles of Frontal Lobe Function*, Oxford University Press, New York, pp51–84

Aron, A. R., Dowson, J. H., Sahakian, B. J. and Robbins, T. W. (2003) 'Methylphenidate improves response inhibition in adults with attention-deficit/hyperactivity disorder', *Biological Psychiatry*, vol 54, pp1465–1468

Austad, S. N. (1997) *Why We Age: What Science is Discovering about the Body's Journey Through Life*, John Wiley and Sons, New York

Bailey, C. H., Bartsch, D. and Kandel, E. R. (1996) 'Toward a molecular definition of long-term memory storage', *Proceedings of the National Academy of Sciences of the United States of America*, vol 93, no 24, pp13,445–13,452

Bainbridge, W. S. (1994) 'Semantic differential', in R. E. Asher and J. M. Y. Simpson (eds) *The Encyclopedia of Language and Linguistics*, Pergamon, Oxford, UK, pp3800–3801

Bainbridge, W. S. (2002a) 'A question of immortality', *Analog*, vol 122, no 5, pp40–49

Bainbridge, W. S. (2002b) 'The spaceflight revolution revisited', in S. J. Garber (ed) *Looking Backward, Looking Forward*, NASA, Washington, DC, pp39–64

Bainbridge, W. S. (2003) 'Massive questionnaires for personality capture', *Social Science Computer Review*, vol 21, no 3, pp267–280

Bainbridge, W. S. (2004a) 'The future of the internet: Cultural and individual conceptions', in P. N. Howard and S. Jones (eds) *Society Online: The Internet in Context*, Sage, Thousand Oaks, CA, pp307–324

Bainbridge, W. S. (2004b) 'Progress toward cyberimmortality', in Immortality Institute, *The Scientific Conquest of Death*, Libros En Red, Montevideo, Uraguay, pp107–122

Bainbridge, W. S. (2005) 'The coming conflict between religion and cognitive science', in C. G. Wagner (ed) *Foresight, Innovation, and Strategy: Toward a Wiser Future*, World Future Society, Bethesda, MD, pp75–87

Bainbridge, W. S. (2006) 'Cognitive technologies', in W. S. Bainbridge and M. C. Roco (eds) *Managing Nano-Bio-Info-Cogno Innovations: Converging Technologies in Society*, Springer, Berlin, pp203–226

Bainbridge, W. S. (2007) 'The scientific research potential of virtual worlds', *Science*, vol 317, no 5837, pp472–476

Barch, D. M. (2004) 'Pharmacological manipulation of human working memory', *Psychopharmacology*, vol 174, no 1, pp126–135

Barnes, C. (1991) *Disabled People in Britain and Discrimination: A Case for Anti-Discrimination Legislation*, Hurst, London

Barnes, D. E., Tager, I. B., Satariano, W. A. and Yaffe K. (2004) 'The relationship between literacy and cognition in well-educated elders', *Journals of Gerontology Series A – Biological Sciences and Medical Sciences*, vol 59, no 4, pp390–395

Barzilai, N. et al (2003) 'Unique lipoprotein phenotype and genotype associated with exceptional longevity', *JAMA*, vol 290, pp2030–2040

Bayertz, K. (2003) 'Human nature: How normative might it be?', *Journal of Medicine and Philosophy*, vol 28, no 2, pp131–150

Baylis, N. V. K. (2005) *Learning from Wonderful Lives: Lessons from the Study of Well-Being*, Cambridge Well-Being Books, Cambridge, UK

Beck, U. (1992) *Risk Society. Towards a New Modernity*, Sage, London

Becker, H. S. (1963) *The Outsiders: Studies in the Sociology of Deviance*, Free Press, New York

Bedelbaeva, K., Gourevitch, D., Clark, L., Chen, P., Leferovich, J. M. and Heber-Katz, E. (2004) 'The MRL mouse heart healing response shows donor dominance in allogeneic fetal liver chimeric mice', *Cloning and Stem Cells*, vol 4, pp352–363

Beja-Pereira, A., Luikart, G. et al (2003) 'Gene-culture coevolution between cattle milk protein genes and human lactase genes', *Nature Genetics*, vol 35, pp311–313

Beltrami, A. P., Barlucchi, L., Torella, D., Baker, M., Limana, F., Chimenti, S., Kasahara, H., Rota, M., Musso, E., Urbanek, K., Leri, A., Kajstura, J., Nadal-Ginard, B. and Anversa, P. (2003) 'Adult cardiac stem cells are multipotent and support myocardial regeneration', *Cell*, vol 114, pp763–776

Berry, D., Rose, C., Remo, B. and Brown, D. D. (1998) 'The expression pattern of thyroid hormone genes in remodeling tadpole tissue defines distinct growth and resorption gene expression programs', *Developmental Biology*, vol 203, pp24–35

Blair, C., Gamson, D., Thorne, S. and Baker, D. (2005) 'Rising mean IQ: Cognitive demand of mathematics education for young children, population exposure to formal schooling, and the neurobiology of the prefrontal cortex', *Intelligence*, vol 33, no 1, pp93–106

Blake, W. (1790) 'The argument', in *The Marriage of Heaven and Hell*, initially self-published and subsequently widely reprinted; text available at www.bibliomania.com/0/2/81/197/frameset.html

Blankenhorn, E. P., Troutman, S., Desquenne, C. L., Zhang, X-M. and Heber-Katz, E. (2003) 'Sexually dimorphic genes regulate healing and regeneration in the MRL mice', *Mammalian Genome*, vol 14, pp250–260

Bloom, D. and Canning, D. (2000) 'The health and wealth of nations', *Science*, vol 287, pp1207–1209

Bode, H. R. (1996) 'The interstitial cell lineage of hydra: A stem cell system that arose early in evolution', *Journal of Cell Science*, vol 109, pp1155–1164

Boellner, S. W., Knutson, J. A., Jiang, J. G., Yang, R. and Earl, C. Q. (2006) 'Modafinil-ADHD: Long-term efficacy and safety in children and adolescents with ADHD', American Psychiatric Association Annual Meeting 2006, online abstracts, control number 2690, http://abstractsonline.com/viewer/?mkey=%7B624DA36A%2D6E6A%2D40F5%2DB834%2D8E6AD1BD1013%7D, accessed 13 June 2008

Boire, R. G. (2001) 'On cognitive liberty', *The Journal of Cognitive Liberties*, vol 2, no 1, pp7–22

Borgens, R. B. (1982) 'Mice regrow the tips of the foretoes', *Science*, vol 217, pp747–750

Bostrom, N. (2005) 'In defence of posthuman dignity', *Bioethics*, vol 19, no 3, pp202–214

Bostrom, N. (2006) 'Welcome to a world of exponential change', in P. Miller and J. Wilsdon (eds) *Better Humans? The Politics of Human Enhancement and Life Extension*, DEMOS, London, pp40–50

Boyce, N. (2004) 'Science calls at the deathbed', *U.S. News and World Report*, 12 January 2004, pp50–51

Boyer, P. (2001) *Religion Explained: The Evolutionary Origins of Religious Thought*, Basic Books, New York

Braidwood, R. J. (1952) *The Near East and the Foundations of Civilization*, Oregon State System of Higher Education, Eugene, OR

Branner, A., Stein, R. B. and Normann, E. A. (2001) 'Selective stimulation of a cat sciatic nerve using an array of varying-length micro electrodes', *Journal of Neurophysiology*, vol 54, no 4, pp1585–1594

Breitenstein, C., Wailke, S., Bushuven, S., Kamping, S., Zwitserlood, P., Ringelstein, E. B. and Knecht, S. (2004) 'D-amphetamine boosts language learning independent of its cardiovascular and motor arousing effects', *Neuropsychopharmacology*, vol 29, no 9, pp1704–1714

Brockes, J. P. and Kumar, A. (2005) 'Appendage regeneration in adult vertebrates and implications for regenerative medicine', *Science*, vol 310, no 5756, pp1919–1923

Brophy, B., Smolenski, G. et al (2003) 'Cloned transgenic cattle produce milk with higher levels of beta-casein and kappa-casein', *Nature Biotechnology*, vol 21, pp157–162

Brown-Borg, H. M. et al (1996) 'Dwarf mice and the ageing process', *Nature*, vol 384, p33

Buchanan, A., Brock, D. W., Daniels, N. and Wikler, D. (2000) *From Chance to Choice*, Cambridge University Press, Cambridge, UK

Buller, D. (2005) *Adapting Minds: Evolutionary Psychology and the Persistent Quest for Human Nature*, MIT Press, Cambridge, MA

Burton, O. V. (ed) (2002) *Computing in the Social Sciences and Humanities*, University of Illinois Press, Urbana, IL

Butefisch, C. M., Khurana, V., Kopylev, L. and Cohen, L. G. (2004) 'Enhancing encoding of a motor memory in the primary motor cortex by cortical stimulation', *Journal of Neurophysiology*, vol 91, no 5, pp2110–2116

Butler, R. N. and Brody, J. A. (eds) (1995) *Delaying the Onset of Late-Life Dysfunction*, Springer Publishing Company, New York

Butler, R. N. et al (2004) 'The aging factor in health and disease: The promise of basic research on aging', Special Report, *Aging Clinical and Experimental Research*, vol 16, pp104–112

Butz, A., Hollerer, T., Feiner, S., McIntyre, B. and Beshers, C. (1999) 'Enveloping users and computers in a collaborative 3D augmented reality', *International Workshop on Augmented Reality*, no 99, 20–21 October 1999, San Francisco, CA, pp35–44

Buzan, T. (1982) *Use Your Head*, BBC Books, London

Calef, T., Pieper, M. and Coffey, B. (1999) 'Comparisons of eye movements before and after a speed-reading course', *Journal of the American Optometric Association*, vol 70, no 3, pp171–181

Caplan, A. (2006) 'Is it wrong to try to improve human nature?', in P. Miller and J. Wilsdon (eds) *Better Humans? The Politics of Human Enhancement and Life Extension*, DEMOS, London, pp31–39

Cardenas, D. D., McLean, A., Jr., Farrell-Roberts, L., Baker, L., Brooke, M. and Haselkorn, J. (1994) 'Oral physostigmine and impaired memory in adults with brain injury', *Brain Injury*, vol 8, pp579–587

Carlson, M. R., Komine, Y., Bryant, S. V. and Gardiner, D. M. (2001) 'Expression of Hoxb13 and Hoxc10 in developing and regenerating axolotl limbs and tails', *Developmental Biology*, vol 229, pp396–406

Carmena, J. M., Lebedev, M. A., Crist, R. E., O'Doherty, J. E., Santucci, D. M., Dimitrov, D. F., Patil, P. G., Henriquez, C. S. and Nicolelis, M. A. L. (2003) 'Learning to control a brain–machine interface for reaching and grasping by primates', *PLoS Biology*, vol 1, no 2, pp193–208

Carpenter, M. D., Winsberg, B. G. and Camus, L. A. (1992) 'Methylphenidate augmentation therapy in schizophrenia', *Journal of Clinical Psychopharmacology*, vol 12, pp273–275

Carrere, E. (2005) *I am Alive and you are Dead; A Journey into the Mind of Philip K. Dick*, Bloomsbury, London

Cattell, R. (1987) *Intelligence: Its Structure, Growth, and Action*, Elsevier Science, New York

Cayley, D. (1992) *Ivan Illich in Conversation*, House of Anansi Press, Toronto, Canada

Chernoff, E. A. G., O'Hara, C. M., Bauerle, B. and Bowling, M. (2000) 'Matrix metalloproteinase production in regenerating axolotl spinal cord', *Wound Repair and Regeneration*, vol 8, pp282–291

Cherry, S. (2005) 'Total recall', *IEEE Spectrum*, vol 42, no 11, pp24–30

Chinese Ministry of Health (2005) www.moh.gov.cn/public/open.aspx?n_id=9713&seq=按类索引, accessed 13 June 2008

Clark, A. (2002) *Natural Born Cyborgs*, Oxford University Press, Oxford

Coale, A. J. (1973) 'The demographic transition', in International Union for the Scientific Study of Population, *Proceedings of the International Population Conference* (volume 1), Liege, Belgium

Cochran, G., Hardy, J. and Harpending, H. (2006) 'Natural history of Ashkenazi intelligence', *Journal of Biosocial Science*, vol 38, no 5, pp659–694

Cochrane, P. (1997) *Tips for the Time Traveller*, Orion Books, London

Cohen, M., Herder, J. and Martens, W. (1999) 'Cyberspatial audio technology', *Journal of the Acoustical Society of Japan (English)*, vol 20, no 6, pp389–395

Collingridge, D. (1980) *The Social Control of Technology*, Frances Pinter, London

Craig, I. and Plomin, R. (2006) 'Quantitative trait loci for IQ and other complex traits: Single-nucleotide polymorphism genotyping using pooled DNA and microarrays', *Genes, Brain and Behavior*, vol 5, pp32–37

David, C. N. and Campbell, R. D. (1972) 'Cell cycle kinetics and development of *Hydra attenuata* – Epithelial cells', *Journal of Cell Science*, vol 11, pp557–568

Davis, K. (1991) 'Population and resources: Fact and interpretation', in K. Davis and M. S. Bernstam (eds) *Resources, Environment and Population: Present Knowledge*, Oxford University Press, Oxford, UK, pp1–21

Davis, K. (1954) 'The world demographic transition', *Annals of the American Academy of Political and Social Science*, vol 237, pp1–11

Denislic, M. and Meh, D. (1994) 'Neurophysiological assessment of peripheral neuropathy in primary Sjögren's syndrome', *Clinical Investigator*, vol 72, no 11, pp822–829

Dennett, D. (1987) *The Intentional Stance*, Bradford Books, Cambridge, MA

Dennett, D. (1991) *Consciousness Explained*, Little, Brown and Co., London

de Grey A. D. N. J. (2004) 'Escape velocity: Why the prospect of extreme human life extension matters now', *PLoS Biology*, vol 2, no 6, pp723–726

de Grey A. D. N. J. (2005a) 'Resistance to debate on how to postpone ageing is delaying progress and costing lives', *EMBO Reports*, vol 6 (S1), ppS49–S53 de Grey A. D. N. J. (2005b) 'Like it or not, life extension research extends beyond biogerontology', *EMBO Reports*, vol 6, no 11, p1000

de Grey, A. D. N. J., Ames, B. N., Andersen, J. K., Bartke, A., Campisi, J., Heward, C. B., McCarter, R. J. M. and Stock, G. (2002) 'Time to talk SENS: Critiquing the immutability of human aging', *Annals of the New York Academy of Sciences*, vol 959, pp452–462

de Quervain, D. J. F. and Papassotiropoulos, A. (2006) 'Identification of a genetic cluster influencing memory performance and hippocampal activity in humans', *Proceedings of the National Academy of Sciences of the United States of America*, vol 103, no 11, pp4270–4274

Despouy, L. (1993) *Human Rights and Disabled Persons*, UN, New York

Diamond, J. and Bellwood, P. (2003) 'Farmers and their languages: The first expansions', *Science*, vol 300, pp597–603

Diamond, M. C., Johnson, R. E. and Ingham, C. A. (1975) 'Morphological changes in young, adult and aging rat cerebral-cortex, hippocampus, and diencephalon', *Behavioral Biology*, vol 14, no 2, pp163–174

Douglas, B. S. (1972) 'Conservative management of guillotine amputations of the fingers of children', *Australian Paediatric Journal*, vol 8, pp86–90

Douglas, M. (1966) *Purity and Danger: An Analysis of Concepts of Pollution and Taboo*, Routledge and Kegan Paul, London

DPI Europe (2000) 'Position statement on bioethics and human rights', Disabled Peoples International Europe, www.dpi-europe.org/_Media/bioethics-english.pdf, accessed 13 November 2008

Drexler, K. E. (1991) 'Hypertext publishing and the evolution of knowledge', *Social Intelligence*, vol 1, no 2, pp87–120

Dyson, F. J. (1979) 'Time without end: Physics and biology in an open universe', *Reviews of Modern Physics*, vol 51, pp447–460

Echeverri, K., Clarke, J. D. and Tanaka, E. M. (2001) 'In vivo imaging indicates muscle fiber dedifferentiation is a major contributor to the regenerating tail blastema', *Developmental Biology*, vol 236, pp151–164

Edelhoff, S., Villacres, E. C., Storm, D. R. and Disteche, C. M. (1995) 'Mapping of adenylyl-cyclase genes Type-I, Type-Ii, Type-Iii, Type-Iv, Type-V and Type-Vi in mouse', *Mammalian Genome*, vol 6, no 2, pp111–113

Ehrlich, P. (1968) *The Population Bomb*, Sierra Club-Ballantine, New York

Eisenberg, D. M., Kessler, R. C., Foster, C., Norlock, F. E., Calkins, D. R. and Delbanco, T. L. (1993) 'Unconventional medicine in the United States: Prevalence, costs, and patterns of use', *New England Journal of Medicine*, vol 328, pp246–252

Elliott, R., Sahakian, B. J., Matthews, K., Bannerjea, A., Rimmer, J. and Robbins, T. W. (1997) 'Effects of methylphenidate on spatial working memory and planning in healthy young adults', *Psychopharmacology (Berl)*, vol 131, pp196–206

Encarta (2006) 'Distribution of IQ scores', available from http://encarta.msn.com/media_461540296_761570026_-1_1/Distribution_of_IQ_Scores.html, accessed 30 August 2006

Engelbart, D. C. (1962) 'Augmenting human intellect: A conceptual framework', Summary Report AFOSR-3223 under Contract AF 49(638)-1024, SRI Project 3578 for Air Force Office of Scientific Research, Stanford Research Institute, Menlo Park, CA, available from www.bootstrap.org/augdocs/friedewald030402/augmentinghumanintellect/1introduction.html, accessed 16 June 2008

Ericsson, A. K. (2003) 'Exceptional memorizers: Made, not born', *Trends in Cognitive Sciences*, vol 7, no 6, pp233–235

Ericsson, K. A., Chase, W. G. and Faloon, S. (1980) 'Acquisition of a memory skill', *Science*, vol 208, no 4448, pp1181–1182

European Commission (2004) *Converging Technologies: Shaping the Future of European Societies*, Office for Official Publications of the European Communities, Luxemburg

Evenden, J. L. (1999) 'Varieties of impulsivity', *Psychopharmacology (Berl)*, vol 146, pp348–361

Fagan, B. M. (1996) *World Prehistory : A Brief Introduction*, HarperCollins College Publishers, New York

Fan, X., Sun, S., Yen, J., Sun, B., Airy, G., McNeese, M., Hanratty, T. and Dumer, J. (2005a) 'Collaborative RPD-enabled agents assisting the three-block challenge in C2CUT', paper read at 2005 Conference on Behavior Representation in Modeling and Simulation (BRIMS 2005)

Fan, X., Sun, S., McNeese, M. and Yen, J. (2005b) 'Extending the recognition-primed decision model to support human–agent collaboration', paper read at Conference on Autonomous Agents and Multi-Agent Systems (AAMAS), 25–29 July, Utrecht, The Netherlands

Farah, M. J., Illes, J., Cook-Deegan, R., Gardner, H., Kandel, E., King, P., Parens, E., Sahakian, B. and Wolpe, P. R. (2004) 'Neurocognitive enhancement: What can we do and what should we do?', *Nature Reviews Euroscience*, vol 5, no 5, pp421–425

Farrand, P., Hussain, F. and Hennessy, E. (2002) 'The efficacy of the "mind map" study technique', *Medical Education*, vol 36, no 5, pp426–431

Fedoroff, N. V. (2003) 'Agriculture: Prehistoric GM corn', *Science*, vol 302, pp1158–1159

Feltz, D. L. and Landers, D. M. (1983) 'The effects of mental practice on motor skill learning and performance: A meta-analysis', *Journal of Sport Psychology*, vol 5, no 1, pp25–57

Ferrari, G., Cusella-De Angelis, G., Coletta, M., Paolucci, E., Stornaiuolo, A., Cossu, G. and Mavilio, F. (1998) 'Muscle regeneration by bone marrow-derived myogenic progenitors', *Science*, vol 279, pp1528–1530

Ferretti, P., Fekete, D. M., Patterson, M. and Lane, E. B. (1989) 'Transient expression of simple epithelial keratins by mesenchymal cells of regenerating newt limb', *Developmental Biology*, vol 133, pp415–424

Finn, W. and LoPresti, P. (eds) (2003) *Handbook of Neuroprosthetic Methods*, CRC Press, London and Boca Raton, FL

Fletcher, A. (1995) *Obstacles to Overcoming the Integration of Disabled People*, Disability Awareness in Action (DAA), London

Flurkey, K. et al (2001) 'Lifespan extension and delayed immune and collagen aging in mutant mice with defects in growth hormone production', *Proceedings of the National Academy of Sciences*, vol 98, pp6736–6741

Flynn, J. R. (1987) 'Massive IQ gains in 14 nations: What IQ tests really measure', *Psychological Bulletin*, vol 101, no 2, pp171–191

Foresight (2005) 'Brain science, addiction and drugs', www.foresight.gov.uk/OurWork/CompletedProjects/Brain%20Science/index.asp, accessed 16 June 2008

Foster, J. K., Lidder, P. G. and Sunram, S. I. (1998) 'Glucose and memory: Fractionation of enhancement effects?', *Psychopharmacology*, vol 137, no 3, pp259–270

Franklin, S. (2006) 'Better by design?', in P. Miller and J. Wilsdon (eds) (2006) *Better Humans? The Politics of Human Enhancement and Life Extension*, Demos, London, pp86–94

Fregni, F., Boggio, P. S., Nitsche, M., Bermpohl, F., Antal, A., Feredoes, E., Marcolin, M. A., Rigonatti, S. P., Silva, M. T. A., Paulus, W. and Pascual-Leone, A. (2005) 'Anodal transcranial direct current stimulation of prefrontal cortex enhances working memory', *Experimental Brain Research*, vol 166, no 1, pp23–30

Freo, U., Ricciardi, E., Pietrini, P., Schapiro, M. B., Rapoport, S. I. and Furey, M. L. (2005) 'Pharmacological modulation of prefrontal cortical activity during a working memory task in young and older humans: A PET study with physostigmine', *American Journal of Psychiatry*, vol 162, no 11, pp2061–2070

Fukuyama, F. (2002) *Our Posthuman Future: Consequences of the Biotechnology Revolution*, Farrar, Strauss and Giroux, New York

Fukuyama, F. (2004) 'Transhumanism', *Foreign Policy*, no 144, September/October, pp42–43

Gallagher, H. (1995) *By Trust Betrayed: Patients, Physicians and the Licence to Kill in the Third Reich*, Vandamere Press, New York

Galliot, B. and Schmid, V. (2002) 'Cnidarians as a model system for understanding evolution and regeneration', *International Journal for Developmental Biology*, vol 46, pp39–48

Garud, R. and Ahlstrom, D. (1997) 'Technology assessment: A socio-cognitive perspective', *Journal of Engineering and Technology Management*, vol 14, pp25–48

Gasson, M., Hutt, B., Goodhew, I., Kyberd, P. and Warwick, K. (2005) 'Invasive neural prosthesis for neural signal detection and nerve stimulation', *International Journal of Adaptive Control and Signal Processing*, vol 19, no 5, pp365–375

Geesaman B. J. et al (2003) 'Haplotype-based identification of a microsomal transfer protein marker associated with the human lifespan', *Proceedings of the National Academy of Sciences*, vol 100, pp14,115–14,120

Gemmell, J., Williams, L., Wood, K., Bell, G. and Lueder, R. (2004) 'Passive capture and ensuing issues for a personal lifetime store', *Proceedings of the First ACM Workshop on Continuous Archival and Retrieval of Personal Experiences*, Association for Computing Machinery, New York, pp48–55

Geyer, M. A. and Tamminga, C. A. (2004) 'Measurement and treatment research to improve cognition in schizophrenia: Neuropharmacological aspects', *Psychopharmacology*, vol 174, pp1–2

Giles, J. (2005) 'Internet encyclopaedias go head to head', *Nature*, vol 438, no 7070, pp900–901

Gladstone, D. J. and Black, S. E. (2000) 'Enhancing recovery after stroke with noradrenergic pharmacotherapy: A new frontier?', *Canadian Journal of Neurological Sciences*, vol 27, no 2, pp97–105

Globus, M., Vethamany-Globus, S. and Lee, Y. C. I. (1980) 'Effect of apical epidermal cap on mitotic cycle and cartilage differentiation in regeneration blastemata in the newt *Notophthalmus viridescens*', *Developmental Biology*, vol 75, pp358–372

Gonzalez-Estevez, C. and Salo, E. (2001) 'GtDap-1: A molecular marker to follow apoptosis in planarian regeneration', *International Journal for Developmental Biology*, vol 45, S1, pS180

Goss, R. J. (1969) *Principles of Regeneration*, Academic Press, New York

Goss, R. J. and Grimes, L. N. (1975) 'Epidermal downgrowths in regenerating rabbit ear holes', *Journal of Morphology*, vol 146, pp533–542

Gottfredson, L. S. (1997) 'Why g matters: The complexity of everyday life', *Intelligence*, vol 24, no 1, pp79–132

Gottfredson, L. S. (2004) 'Life, death, and intelligence', *Journal of Cognitive Education and Psychology*, vol 4, no 1, pp23–46

Gourevitch, D., Clark, L., Chen, P., Seitz, A., Samulewicz, S. and Heber-Katz, E. (2002) 'Matrix metalloproteinase activity correlates with blastema formation in the regenerating MRL ear hole model', *Developmental Dynamics*, vol 226, pp377–387

Governale, L. and Kaplan, S. (2005) 'One year post-pediatric exclusivity postmarketing adverse event review: Drug use data', Food and Drug Administration, US Department of Health and Human Services, Memorandum PID# D040058, available at www.fda.gov/ohrms/dockets/ac/05/briefing/2005-4152b2_01_02_Concerta%20Use%20Cleared.pdf, accessed 16 June 2008

Greenfield, S. (2003) *Tomorrow's People: How 21st Century Technology is Changing the Way we Think and Feel*, Allen Lane, London

Greenhill, L. L., Biederman, J., Boellner, S., Rugino, T. A., Sangal, R. B. and Swanson, J. M. (2005) 'Modafinil film-coated tablets significantly improve symptoms on ADHD Rating Scale-IV School and Home and overall clinical condition in children and adolescents with attention-deficit/hyperactivity disorder', New Clinical Drug Evaluation Unit (NCDEU), 45th Annual Meeting, Boca Raton, FL, Session I-88

Greenough, W. T. and Volkmar, F. R. (1973) 'Pattern of dendritic branching in occipital cortex of rats reared in complex environments', *Experimental Neurology*, vol 40, no 2, pp491–504

Grillo, H. C., Lapiere, C. M., Dresden, M. H. and Gross, J. (1968) 'Collagenolytic activity in regenerating forelimbs of the adult newt', *Developmental Biology*, vol 17, pp571–583

Grunwald, A. (2006) 'Nanotechnologie als Chiffre der Zukunft', in A. Nordmann, J. Schummer and A. Schwarz (eds) *Nanotechnologien im Kontext: Philosophische, Ethische, Gesellschaftliche Perspektiven*, Akademische Verlagsgesellschaft, Berlin, pp49–80

Gulpinar, M. A. and Yegen, B. C. (2004) 'The physiology of learning and memory: Role of peptides and stress', *Current Protein and Peptide Science*, vol 5, no 6, pp457–473

Han, M., Yang, X., Farrington, J. E. and Muneoka, K. (2003) 'Digit regeneration is regulated by Msx1 and BMP4 in fetal mice', *Development*, vol 130, pp5123–5132

Hanson, R., 'the economics of science fiction', http://hanson.gmu.edu/econofsf.html, accessed 16 June 2008

Hanson, R., Opre, R. and Porter, D. (2006) 'Information aggregation and manipulation in an experimental market', *Journal of Economic Behavior and Organization*, vol 60, no 4, pp449–459

Hanson, R., Polk, C., Ledyard, J. and Ishikida, T. (2003) 'The policy analysis market: An electronic commerce application of a combinatorial information market', paper read at Association for Computing Machinery (ACM) Conference on Electronic Commerce

Hare, B., Brown, M. et al (2002) 'The domestication of social cognition in dogs', *Science*, vol 298, no 5598, pp1634–1636

Harper, S. (2004) 'The challenge of families and demographic ageing', in S. Harper (ed) *Families in Ageing Societies: A multidisciplinary approach*, Oxford University Press, Oxford, UK

Harris, J. (2007) *Enhancing Evolution: The Ethical Case for Making Better People*, Princeton University Press, Princeton, NJ

He, G. et al (2002) 'An analysis on immunization coverage rate of children in different economic development areas of Gansu Province', *Immunization in China*, vol 8, no 4

Head, M. (2006) 'Thanks but I'd rather be disabled', *Listener – The Things that Matter*, vol 202, no 3428, 21 January, pp48–49, available at www.listener.co.nz/issue3428/5313.html, accessed 16 June 2008

Healey, J. and Picard, R. W. (1998) 'StartleCam: A cybernetic wearable camera', paper read at Second International Symposium on Wearable Computing, Pittsburgh, PA

Heber-Katz, E. (1999) 'The regenerating mouse ear', *Seminars in Cell and Developmental Biology*, vol 10, no 4, pp415–419

Heber-Katz, E. (2005) 'Methods and compositions for healing heart wounds', US Patent No 6,852,706

Heber-Katz, E., Leferovitch, J., Bedelbaeva, K. and Gourevitch, D. (2004a) 'Spallanzani's mouse: A model of restoration and regeneration', *Current Topics in Microbiology and Immunology*, vol 280, pp165–189

Heber-Katz, E., Chen, P., Clark, L., Zhang, X-M., Troutman, S. and Blankenhorn, E. P. (2004b) 'Regeneration in MRL mice: Further genetic loci controlling the ear hole closure trait using MRL and M.m. Castaneus mice', *Wound Repair and Regeneration*, vol 12, pp384–392

Heise, D. R. (1970) 'The semantic differential and attitude research', in G. F. Summers (ed) *Attitude Measurement*, Rand McNally, Chicago, IL, pp235–253

Heise, D. R. and Weir, B. (1999) 'A test of symbolic interactionist predictions about emotions in imagined situations', *Symbolic Interaction*, vol 22, pp129–161

Helland, I. B., Smith, L., Saarem, K., Saugstad, O. D. and Drevon, C. A. (2003) 'Maternal supplementation with very-long-chain n-3 fatty acids during pregnancy and lactation augments children's IQ at 4 years of age', *Pediatrics*, vol 111, no 1, pp39–44

HLEG (2004) 'Foresighting the new technology wave', in *Converging Technologies: Shaping the Future of European Societies*, High Level Expert Group, Office for Official Publications of the European Communities, Luxemburg

Hofmann, S. G., Meuret, A. E., Smits, J. A. J., Simon, N. M., Pollack, M. H., Eisenmenger, K., Shiekh, M. and Otto, M. W. (2006) 'Augmentation of exposure therapy with D-cycloserine for social anxiety disorder', *Archives of General Psychiatry*, vol 63, no 3, pp298–304

Hockey, R. (1973) 'Changes in information-selection patterns in multisource monitoring as a function of induced arousal shifts', *Journal of Experimental Psychology*, vol 101, pp35–42

Holm, S. (1998) 'A life in the shadow: One reason why we should not clone humans', *Cambridge Quarterly of Healthcare Ethics*, vol 7, no 4, pp417–435

Holsboer, F. (1999) 'The rationale for corticotrophin-releasing hormone receptor (CRH-R) antagonists to treat depression and anxiety', *Journal of Psychiatric Research*, vol 33, pp181–214

Holstein, T. W., Hobmayer, E. and Technau, U. (2003) 'Cnidarians: An evolutionarily conserved model system for regeneration?', *Developmental Dynamics*, vol 226, pp257–267

House of Lords Science and Technology Committee (2005) 'Ageing: Scientific aspects', HL Paper 20-1, The Stationery Office, London, available at www.publications.parliament. uk/pa/ld200506/ldselect/ldsctech/20/20i.pdf, accessed 17 June 2008

Hughes, J. (2004) 'Battle plan to be more than well: Transhumanism is finally getting in gear', http://transhumanism.org/index.php/th/more/509/, accessed 17 June 2008

Hughes, T. P. (1986) 'The seamless web: Technology, science, etcetera, etcetera', *Social Studies of Science*, vol 16, pp281–292

Hummel, F. C. and Cohen, L. G. (2005) 'Drivers of brain plasticity', *Current Opinion in Neurology*, vol 18, no 6, pp667–674

Huppert, F., Baylis, N. and Keverne, B. (eds) (2005) *The Science of Well-Being*, Oxford University Press, Oxford, UK

Hurst, R. (2006) 'The perfect crime', in P. Miller and J. Wilsdon (eds) (2006) *Better Humans? The Politics of Human Enhancement and Life Extension*, Demos, London, pp114–121

Illingworth, C. M. (1974) 'Trapped fingers and amputated finger tips in children', *Journal of Pediatric Surgery*, vol 9, no 6, pp853–858

Ingvar, M., AmbrosIngerson, J., Davis, M., Granger, R., Kessler, M., Rogers, G. A., Schehr, R. S. and Lynch, G. (1997) 'Enhancement by an ampakine of memory encoding in humans', *Experimental Neurology*, vol 146, no 2, pp553–559

Institute of Nanotechnology (2005) 'Research applications and markets in nanotechnology in Europe 2005, www.researchandmarkets.com/reportinfo.asp?report_id=302091&t=t&cat_id=4

Iversen, S. D. (1998) 'The pharmacology of memory', *Comptes Rendus de l'Academie des Sciences: Serie III, Sciences de la Vie*, vol 321, nos 2–3, pp209–215

Jackson, P. L., Doyon, J., Richards, C. L. and Malouin, F. (2004) 'The efficacy of combined physical and mental practice in the learning of a foot-sequence task after stroke: A case report', *Neurorehabilitation and Neural Repair*, vol 18, no 2, pp106–111

James, C. (2004) 'Preview: Global status of commercialized biotech/GM crops', ISAAA Briefs No 32, International Service for the Acquisition of Agri-biotech Applications, Ithaca, NY

Jebara, T., Eyster, C., Weaver, J., Starner, T. and Pentland, A. (1997) 'Stochasticks: Augmenting the billiards experience with probabilistic vision and wearable computers', paper read at the International Symposium on Wearable Computers, Cambridge, MA

Johnston, G. (2004) 'Healthy, wealthy and wise? A review of the wider benefits of education', New Zealand Treasury Working Paper 04/04, Wellington, New Zealand

Kaczynski, T. (1995) 'Unabomber Manifesto', http://cyber.eserver.org/unabom.txt, accessed 17 June 2008

Kallen, E. (2004) *Social Inequality and Social Injustice: A Human Rights Perspective*, Macmillan Palgrave, Basingstoke, UK, and New York

Kanda, H., Yogi, T., Ito, Y., Tanaka, S., Watanabe, M. and Uchikawa, Y. (1999) 'Efficient stimulation inducing neural activity in a retinal implant', *Proceedings of IEEE International Conference on Systems, Man and Cybernetics*, vol 4, pp409–413

Kapner, E. (2003) 'Recreational use of Ritalin on college campuses', InfoFactsResources – The Higher Education Center for Alcohol and Other Drug Prevention, www.edc.org/hec/pubs/factsheets/ritalin.pdf, accessed 17 June 2008

Karatzas, C. N. (2003) 'Designer milk from transgenic clones', *Nature Biotechnology*, vol 21, pp138–139

Kass, L. (2003) 'Ageless bodies, happy souls: Biotechnology and the pursuit of perfection', *The New Atlantis*, vol 1

Keck, M., Welt, T., Muller, M. et al (2002) 'Repetitive transcranial magnetic stimulation increases the release of dopamine in the mesolimbic and mesostriatal system', *Neuropharmacology*, vol 43, p101–109

Kelly, K. (2006) 'The next 100 years of science', interview with Kevin Kelly by Stewart Brand, the Long Now Foundation, Cowell Theater, San Francisco, CA, 10 March, transcript available at http://fora.tv/fora/fora_transcript_pdf.php?cid=365

Kennedy, D. O., Pace, S., Haskell, C., Okello, E. J., Milne, A. and Scholey, A. B. (2006) 'Effects of cholinesterase inhibiting sage (*Salvia officinalis*) on mood, anxiety and performance on a psychological stressor battery', *Neuropsychopharmacology*, vol 31, no 4, pp845–852

Kennedy, P. R. and Bakay, R. A. E. (1998) 'Restoration of neural output from a paralyzed patient by a direct brain connection', *Neuroreport*, vol 9, no 8, pp1707–1711

Kennedy, P., Bakay, R., Moore, M., Adams, K. and Goldwaithe, J. (2000) 'Direct control of a computer from the human central nervous system', *IEEE Transactions on Rehabilitation Engineering*, vol 8, no 2, pp198–202

Kennedy, P., Andreasen, D., Ehirim, P., King, B., Kirby, T., Mao, H. and Moore, M. (2004) 'Using human extra-cortical local field potentials to control a switch', *Journal of Neural Engineering*, vol 1, no 2, pp72–77

Khan, P., Linkhart, B. and Simon, H. G. (2002) 'Different regulation of T-box genes Tbx4 and Tbx5 during limb development and limb regeneration', *Developmental Biology*, vol 250, pp383–392

Khush, G. S. (2001) 'Green Revolution: The way forward', *Nature Reviews Genetics*, vol 2, pp815–822

Kijas, J. M. and Andersson, L. (2001) 'A phylogenetic study of the origin of the domestic pig estimated from the near-complete mtDNA genome', *Journal of Molecular Evolution*, vol 52, pp302–308

Kincses, T. Z., Antal, A., Nitsche, M. A., Bartfai, O. and Paulus, W. (2004) 'Facilitation of probabilistic classification learning by transcranial direct current stimulation of the prefrontal cortex in the human', *Neuropsychologia*, vol 42, no 1, pp113–117

Kirkwood, T. B. L. (1999) *Time of our Lives: The Science of Human Aging*, Oxford University Press, Oxford, UK, and New York

Kirkwood, T. B. L. (2001) *The End of Age*, BBC, in association with Profile Books, London

Kleinberg, J. M. (1999) 'Authoritative sources in a hyperlinked environment', *Journal of the ACM*, vol 46, no 5, pp604–632

Kloner, R. A. (2006) 'Attempts to recruit stem cells for repair of acute myocardial infarction: A dose of reality', *Journal of the American Medical Association*, vol 295, pp1058–1060

Kobayashi, M., Hutchinson, S., Theoret, H., Schlaug, G. and Pascual-Leone, A. (2004) 'Repetitive TMS of the motor cortex improves ipsilateral sequential simple finger movements', *Neurology*, vol 62, no 1, pp91–98

Korol, D. L. and Gold, P. E. (1998) 'Glucose, memory, and aging', *American Journal of Clinical Nutrition*, vol 67, no 4, pp764s–771s

Kostoff, R., Murday, J., Lau, C. and Tolles, W. (2006) 'The seminal literature of nanotechnology research', *Journal of Nanoparticle Research*, vol 8, no 2, pp193–213

Kubler, A., Kotchoubey, B., Hinterberger, T., Ghanayim, N., Perelmouter, J., Schauer, M., Fritsch, C., Taub, E. and Birbaumer, N. (1999) 'The thought translation device: A neurophysiological approach to communication in total motor paralysis', *Experimental Brain Research*, vol 124, no 2, pp223–232

Kuroiwa, Y., Kasinathan, P. et al (2002) 'Cloned transchromosomic calves producing human immunoglobulin', *Nature Biotechnology*, vol 20, pp889–894

Kurzweil, R. (2005) *The Singularity is Near: When Humans Transcend Biology*, Viking, New York, NY

Lanza, R., Moore, M. A., Wakayama, T., Perry, A. C., Shieh J. H., Hendrikx, J., Leri, A., Chimenti, S., Monsen, A., Nurzynska, D., West, M. D., Kajstura, J. and Anversa, P. (2004) 'Regeneration of the infarcted heart with stem cells derived by nuclear transplantation', *Circulation Research*, vol 94, pp820–827

Lashley, K. S. (1917) 'The effects of strychnine and caffeine upon rate of learning', *Psychobiology*, vol 1, pp141–169

Lee, E. H. Y. and Ma, Y. L. (1995) 'Amphetamine enhances memory retention and facilitates norepinephrine release from the hippocampus in rats', *Brain Research Bulletin*, vol 37, no 4, pp411–416

Leferovich, J., Bedelbaeva, K., Samulewicz, S., Xhang, X-M., Zwas, D. R., Lankford, E. B. and Heber-Katz, E. (2001) 'Heart regeneration in adult MRL mice', *Proceedings of the National Academy of Sciences*, vol 98, pp9830–9835

Leuthardt, E., Schalk, G., Wolpaw, J., Ojemann, J. and Moran, D. (2004) 'A brain–computer interface using electrocorticographic signals in humans', *Journal of Neural Engineering*, vol 1, no 2, pp63–71

Levin, A. (2000) 'Why must my son be left to die?', *Daily Mail*, 3 August

Levin, Y. (2003) 'The paradox of conservative bioethics', *The New Atlantis*, no 1, pp53–65

Levy, D. L., Smith, M., Robinson, D., Jody, D., Lerner, G., Alvir, J., Geisler, S. H., Szymanski, S. R., Gonzalez, A., Mayerhoff, D. I. et al. (1993) 'Methylphenidate increases thought disorder in recent onset schizophrenics, but not in normal controls', *Biological Psychiatry*, vol 34, pp507–514

Liao, L., Fox, D. and Kautz, H. (2005) 'Location-based activity recognition using relational Markov networks', *Proceedings of the Nineteenth International Joint Conference on Artificial Intelligence*, International Joint Conference on Artificial Intelligence, Edinburgh

Lieberman, H. R. (2001) 'The effects of ginseng, ephedrine, and caffeine on cognitive performance, mood and energy', *Nutrition Reviews*, vol 59, no 4, pp91–102

Liesi, E. et al (2003) 'Alzheimer disease in the US population: Prevalence estimates using the 2000 census', *Archives of Neurology*, vol 60, pp1119–1122

Light, R. (2004) *A Short Report of the DAA Violations Database*, Disability Awareness in Action, London

Lingford-Hughes, A. and Nutt, D. (2003) 'Neurobiology of addiction and implications for treatment', *British Journal of Psychiatry*, vol 182, pp97–100

Lutz, W. (2006) 'Global demographic trends, education, and health', in *Interactions Between Global Change and Human Health*, Scripta Varia 106, The Pontifical Academy of Sciences, Vatican City, ISBN 978-88-7761-085-0, pp252–268

Lutz, W. and Goldstein, J. (guest eds) (2004) 'How to deal with uncertainty in population forecasting?', *International Statistical Review*, vol 72, nos 1–2, pp1–106 and 157–208

Lutz, W., Sanderson, W. and Scherbov, S. (1997) 'Doubling of world population unlikely', *Nature*, vol 387, pp803–805

Lutz, W., Sanderson, W. and Scherbov, S. (2001) 'The end of world population growth', *Nature*, vol 412, pp543–545

Lutz, W., Sanderson, W. C. and Scherbov, S. (eds) (2004) *The End of World Population Growth in the 21st Century: New Challenges for Human Capital Formation and Sustainable Development*, Earthscan, London

Lutz, W., Samir, K.C., Khan, H. T. A., Scherbov, S. and Leeson, G. W. (2007) 'Future ageing in Southeast Asia: Demographic trends, human capital and health status', Interim Report IR-07-026, International Institute for Applied Systems Analysis, Laxenburg, Austria

Lynch, G. (1998) 'Memory and the brain: Unexpected chemistries and a new pharmacology', *Neurobiology of Learning and Memory*, vol 70, nos 1–2, pp82–100

Lynch, G. (2002) 'Memory enhancement: The search for mechanism-based drugs', *Nature Neuroscience*, vol 5, pp1035–1038

Mackie, J. (1984) 'Rights, utility, and universalization', in R. G. Frey (ed) *Utility and Right*, University of Minnesota Press, Minneapolis, MN, pp86–105

Mann, S. (1997) 'Wearable computing: A first step toward personal imaging', *Computer*, vol 30, no 2, pp25–31

Mann, S. (2001) 'Wearable computing: Toward humanistic intelligence', *IEEE Intelligent Systems*, vol 16, no 3, pp10–15

Mann, S. and Niedzviecki, H. (2001) *Cyborg: Digital Destiny and Human Possibility in the Age of the Wearable Computer*, Doubleday Canada, Toronto, Canada

Mannheim, K. (1952) *Essays on the Sociology of Knowledge*, Oxford University Press, Oxford, UK

Manton, K. G., Gu, X. and Lamb, V. L. (2006) 'Long-term trends in life expectancy and active life expectancy in the United States', *Population and Development Review*, vol 32, no 1, pp81–106

Marris, C., Rip, A. and Joly, P-B. (2008) 'Interactive technology assessment in the real world: Dual dynamics in an iTA exercise on genetically modified vines', *Science, Technology and Human Values*, vol 33, no 1, pp77–100

Marshall, L., Molle, M., Hallschmid, M. and Born, J. (2004) 'Transcranial direct current stimulation during sleep improves declarative memory', *Journal of Neuroscience*, vol 24, no 44, pp9985–9992

Martin, J. (2007) *The Meaning of the 21st Century: A Vital Blueprint for Ensuring our Future*, Transworld, London

Martinez, D. E. (1998) 'Mortality patterns suggest lack of senescence in hydra', *Experimental Gerontology*, vol 33, pp217–222

Masinde, G. L., Li, X., Gu, W., Davidson, H., Mohan, S. and Baylink, D. J. (2001) 'Identification of wound healing/regeneration quantitative trait loci (QTL) at multiple time points that explain seventy percent of variance in (MRL/MpJ and SJL/J) mice F2 population', *Genome Research*, vol 11, pp2027–2033

Mayberg, H. S., Lozano, A. M., Voon, V., McNeely, H. E., Seminowicz, D., Hamani, C., Schwalb, J. S. and Kennedy, S. H. (2005) 'Deep Brain Stimulation for Treatment-Resistant Depression', *Neuron*, vol 45, pp651–660

McBrearty, B. A., Desquenne-Clark, L., Zhang, X-M., Blankenhorn, E. P. and Heber-Katz, E. (1998) 'Genetic analysis of a mammalian wound healing trait', *Proceedings of the National Academy of Sciences*, vol 95, pp11,792–11,797

McDowell, S., Whyte, J. and D'Esposito, M. (1998) 'Differential effect of a dopaminergic agonist on prefrontal function in traumatic brain injury patients', *Brain*, vol 121, pp1155–1164

McKibben, B. (2003) 'Designer genes', *The Orion*, May/June, available at www.geneticsandsociety.org/article.php?id=153, accessed 17 June 2008

McTavish, S. F., McPherson, M. H., Sharp, T. et al (1999) 'Attenuation of some subjective effects of amphetamine following tyrosine depletion', *Journal of Psychopharmacology*, vol 13, pp144–147

McTavish, S. F., McPherson, M. H., Harmer, C. J. et al (2001) 'Antidopaminergic effects of dietary tyrosine depletion in healthy subjects and patients with manic illness', *British Journal of Psychiatry*, vol 179, pp356–360

Meck, W. H., Smith, R. A. and Williams, C. L. (1988) 'Prenatal and postnatal choline supplementation produces long-term facilitation of spatial memory', *Developmental Psychobiology*, vol 21, no 4, pp339–353

Mehta, M. A., Calloway, P. and Sahakian, B. J. (2000a) 'Amelioration of specific working memory deficits by methylphenidate in a case of adult attention deficit/hyperactivity disorder', *Journal of Psychopharmacology*, vol 14, pp299–302

Mehta, M. A., Owen, A. M., Sahakian, B. J., Mavaddat, N., Pickard, J. D. and Robbins, T. W. (2000b) 'Methylphenidate enhances working memory by modulating discrete frontal and parietal lobe regions in the human brain', *The Journal of Neuroscience*, vol 20, RC65, pp1–6

Mehta, M. A., Goodyer, I. M. and Sahakian, B. J. (2004) 'Methylphenidate improves working memory and set-shifting in AD/HD: Relationships to baseline memory capacity', *Journal of Child Psychology and Psychiatry*, vol 45, pp293–305

Meikle, A., Riby, L. M. and Stollery, B. (2005) 'Memory processing and the glucose facilitation effect: The effects of stimulus difficulty and memory load', *Nutritional Neuroscience*, vol 8, no 4, pp227–232

Mellott, T. J., Williams, C. L., Meck, W. H. and Blusztajn, J. K. (2004) 'Prenatal choline supplementation advances hippocampal development and enhances MAPK and CREB activation', *Journal of the Federation of American Societies for Experimental Biology*, vol 18, no 1, pp545–547

Metts, R. L. (2001) 'The fatal flaw in the disability adjusted life year', *Disability and Society*, vol 16, no 3

Miller, R. A. (1997) 'When will the biology of aging become useful? Future landmarks in biomedical gerontology', *Journal of the American Geriatrics Society*, vol 45, pp1258–1267

Miller, R. A. (2002) 'Extending life: Scientific prospects and political obstacles', *The Millbank Quarterly*, vol 80, no 1, pp155–174

Miller, R. A. (2003) 'The biology of aging and longevity', in W. R. Hazzard, J. P. Blass, J. B. Halter, J. G. Ouslander and M. E. Tinetti (eds) *Principles of Geriatric Medicine and Gerontology*, McGraw Hill, Inc., New York, pp3–15

Miller, R. A. and Austad, S. N. (2006) 'Growth and aging: Why do big dogs die young?', in E. J. Masoro and S. N. Austad (eds) *Handbook of the Biology of Aging*, Academic Press, New York, pp512–533

Miller, P. and Wilsdon, J. (eds) (2006) *Better Humans? The Politics of Human Enhancement and Life Extension*, Demos, London

Miyazaki, K., Uchiyawa, K., Imokawa, Y. and Yoshizato, K. (1996) 'Cloning and characterization of cDNAs for matrix metalloproteinases of regenerating newt limbs', *Proceedings of the National Academy of Sciences*, vol 93, pp 6819–6824

Mondadori, C. (1996) 'Nootropics: Preclinical results in the light of clinical effects; Comparison with tacrine', *Critical Reviews in Neurobiology*, vol 10, nos 3–4, pp357–370

Morris, E. (2003) *The Fog of War: Eleven Lessons from the Life of Robert McNamara*, DVD, Sony Pictures, US

Morrison, J. L., Loof, S., Pingping, H. and Simon, A. (2006) 'Salamander limb regeneration involves the activation of a multipotent skeletal muscle satellite cell population', *Journal of Cell Biology*, vol 172, pp433–440

Muir, T. and Zegarac, M. (2001) 'Societal costs of exposure to toxic substances: Economic and health costs of four case studies that are candidates for environmental causation', *Environmental Health Perspectives*, vol 109, pp885–903

Muller, U., Steffenhagen, N., Regenthal, R. and Bublak, P. (2004) 'Effects of modafinil on working memory processes in humans', *Psychopharmacology*, vol 177, nos 1–2, pp161–169

Murray, C. J. and Acharya, A. K. (1997) 'Understanding DALYs (disability-adjusted life years)', *Journal of Health Economics*, vol 16, no 6, pp703–730

Nava, E., Landau, D., Brody, S., Linder, L. and Schachinger, H. (2004) 'Mental relaxation improves long-term incidental visual memory', *Neurobiology of Learning and Memory*, vol 81, no 3, pp167–171

Neisser, U. (1997) 'Rising scores on intelligence tests', *American Scientist*, vol 85, no 5, pp440–447

Newhouse, P. A., Potter, A. and Singh, A. (2004) 'Effects of nicotinic stimulation on cognitive performance', *Current Opinion in Pharmacology*, vol 4, no 1, pp36–46

Newmark, P. A. and Sanchez Alvarado, A. (2000) 'Bromodeoxyuridine specifically labels the regenerative stem cells of planarians', *Developmental Biology*, vol 220, p142–153

Nicolelis, M. A. L., Dimitrov, D., Carmena, J. M., Crist, R., Lehew, G., Kralik, J. D. and Wise, S. P. (2003) 'Chronic, multisite, multielectrode recordings in macaque monkeys', *Proceedings of the National Academy of Sciences of the USA*, vol 100, no 19, pp11,041–11,046

NIDA InfoFacts (2005) 'Methylphenidate (Ritalin)', National Institute on Drug Abuse, www.drugabuse.gov/pdf/infofacts/Ritalin05.pdf, accessed 19 June 2008

Nilsson, M., Perfilieva, E., Johansson, U., Orwar, O. and Eriksson, P. S. (1999) 'Enriched environment increases neurogenesis in the adult rat dentate gyrus and improves spatial memory', *Journal of Neurobiology*, vol 39, no 4, pp569–578

Nitsche, M. A., Schauenburg, A., Lang, N., Liebetanz, D., Exner, C., Paulus, W. and Tergau, F. (2003) 'Facilitation of implicit motor learning by weak transcranial direct current stimulation of the primary motor cortex in the human', *Journal of Cognitive Neuroscience*, vol 15, no 4, pp619–626

Nordmann, A. (2007) 'If and then: A critique of speculative nanoethics', *NanoEthics*, vol 1, pp31–46

Nordmann, A. (2008) 'Ignorance at the heart of science? Incredible narratives of brain–machine interfaces', forthcoming in J. S. Ach and B. Lüttenberg (eds) *Ethics in Nanomedicine*, Lit-Verlag, Berlin

Notestein, F. W. (1945) 'Population: The long view', in T. W. Schultz (ed) *Food for the World*, University of Chicago Press, Chicago, IL, pp36–57

Nutt, D. J. (2003) 'The unhappy saga of "happy pills"', *Journal of Psychopharmacology*, vol 17, p251

Nutt, D. J., Robbins, T. W., Stimson, G. V., Ince, M. and Jackson, A. (2006) *Drugs and the Future: Brain Science, Addiction and Society*, Elsevier, Oxford

Nyberg, L., Sandblom, J., Jones, S., Neely, A. S., Petersson, K. M., Ingvar, M. and Backman, L. (2003) 'Neural correlates of training-related memory improvement in adulthood and aging', *Proceedings of the National Academy of Sciences of the USA*, vol 100, no 23, pp13,728–13,733

Odelberg, S. J., Kollhoff, A. and Keating, M. T. (2000) 'Dedifferentiation of mammalian myotubes induced by msx1', *Cell*, vol 103, pp1099–1109

Oliver, M. (1990) *The Politics of Disablement*, MacMillan, Basingstoke, UK

Olshansky, S. J. (1987) 'Simultaneous/multiple cause delay: An epidemiological approach to projecting mortality', *Journals of Gerontology*, vol 42, pp358–365

Olshansky, S. J. (2003) 'Can we justify efforts to slow the rate of aging in humans?', presentation before the annual meeting of the Gerontological Society of America

Olshansky, S. J., Carnes, B. A. and Cassel, C. (1990) 'In search of Methuselah: Estimating the upper limits to human longevity', *Science*, vol 250, pp634–640

Olshansky, S. J. et al (2002) 'Position statement on human aging', *Journals of Gerontology Series A – Biological and Medical Sciences*, vol 57A, no 8, ppB1–B6

Olshansky, S. J. et al (2005) 'A potential decline in life expectancy in the United States in the 21st century', *New England Journal of Medicine*, vol 352, pp1138–1145

Oppenheimer, T. (2003) *The Flickering Mind*, Random House, New York

Osgood, C. E., Suci, G. J. and Tannenbaum, P. H. (1957) *The Measurement of Meaning*, University of Illinois Press, Urbana, IL

Otto, J. J. and Campbell, R. D. (1977) 'Tissue economics of hydra: Regulation of cell cycle, animal size and development by controlled feeding rates', *Journal of Cell Science*, vol 28, pp117–132

Parfit, D. (1997) 'Equality or priority?', *Ratio*, vol 10, no 3, pp202–221

Parker, H. G., Kim, L. V. et al (2004) 'Genetic structure of the purebred domestic dog', *Science*, vol 304, pp1160–1164

Pascual-Leone, A., Tarazona, F., Keenan, J., Tormos, J. M., Hamilton, R. and Catala, M. D. (1999) 'Transcranial magnetic stimulation and neuroplasticity', *Neuropsychologia*, vol 37, no 2, pp207–217

Patil, P. G., Carmena, L. M., Nicolelis, M. A. L. and Turner, D. A. (2004) 'Ensemble recordings of human subcortical neurons as a source of motor control signals for a brain-machine interface', *Neurosurgery*, vol 55, no 1, pp27–35

Patten, B. M. (1990) 'The history of memory arts', *Neurology*, vol 40, no 2, pp346–352

Patterson, D., Fox, D., Kautz, H. and Philipose, M. (2005) 'Fine-grained activity recognition by aggregating abstract object usage', *Proceedings of the IEEE International Symposium on Wearable Computers*, Osaka, Japan

Penny, W., Roberts, S., Curran, E. and Stokes, M. (2000) 'EEG-based communication: A pattern recognition approach', *IEEE Transactions on Rehabilitation Engineering*, vol 8, no 2, pp214–215

Persaud, R. (2006) 'Does smart mean happier?', in P. Miller and J. Wilsdon (eds) *Better Humans, The Politics of Human Enhancement and Life Extension*, Demos, London

Peterman, M. C., Noolandi, J., Blumenkranz, M. S. and Fishman, H. A. (2004) 'Localized chemical release from an artificial synapse chip', *Proceedings of the National Academy of Sciences of the USA*, vol 101, no 27, pp9951–9954

Petersson, K. M., Reis, A., Askelof, S., Castro-Caldas, A. and Ingvar, M. (2000) 'Language processing modulated by literacy: A network analysis of verbal repetition in literate and illiterate subjects', *Journal of Cognitive Neuroscience*, vol 12, no 3, pp364–382

Philips, M. (2006) 'Many human genes evolved recently', *New Scientist*, 7 March, www.newscientist.com/channel/being-human/dn8812.html, accessed 18 June 2008

Picard, R. W. (1997) *Affective Computing*, IT Press, Cambridge, MA

Pinker, S. (2003) 'Can we change human nature?', talk presented at 'The Future of Human Nature' symposium, Frederick S. Pardee Center for the Study of the Longer-Range Future, Boston University, Boston, MA, 11–12 April

Pitman, R. K., Sanders, K. M., Zusman, R. M., Healy, A. R., Cheema, F., Lasko, N. B., Cahill, L. and Orr, S. P. (2002) 'Pilot study of secondary prevention of posttraumatic stress disorder with propranolol', *Biological Psychiatry*, vol 51, no 2, pp189–192

Poboroniuc, M. S., Fuhr, T., Riener, R. and Donaldson, N. (2002) 'Closed-loop control for FES-supported standing up and sitting down', in *Proceedings of the 7th Conference of the International Functional Electrical Stimulation Society (IFESS)*, Ljubljana, Slovenia, pp307–309

Pollack, M. E. (2005) 'Intelligent technology for an aging population: The use of AI to assist elders with cognitive impairment', *AI Magazine*, vol 26, no 2, pp9–24

Popovic, M. R., Keller, T., Moran, M. and Dietz, V. (1998) 'Neural prosthesis for spinal cord injured subjects', *Journal Bioworld*, vol 1, no 1, pp6–9

Power, A. E., Vazdarjanova, A. and McGaugh, J. L. (2003) 'Muscarinic cholinergic influences in memory consolidation', *Neurobiology of Learning and Memory*, vol 80, no 3, pp178–193

Price, J. S., Allen, S., Faucheux, C., Althnaian, T. and Mount, J. G. (2005) 'Deer antlers: A zoological curiosity or the key to understanding organ regeneration in mammals?', *Journal of Anatomy*, vol 207, pp603–618

Public Agenda (2005) 'The science of aging gracefully: Scientists and the public talk about aging research', The Alliance for Aging Research and the American Federation for Aging Research, Washington, DC

Putnam, R. D. (2000) *Bowling Alone: The Collapse and Revival of American Community*, Simon and Schuster, New York

Quinones, J. L., Rosa, R., Ruiz, D. L. and Garcia-Arraras, J. E. (2002) 'Extracellular matrix remodeling and metalloproteinase involvement during intestine regeneration in the sea cucumber *Holothuria glaberrima*', *Developmental Biology*, vol 250, pp181–197

Rahman, S., Robbins, T. W., Hodges, J. R., Mehta, M. A., Nestor, P. J., Clark, L. and Sahakian, B. J. (2005) 'Methylphenidate ("Ritalin") can ameliorate abnormal risk-taking behaviour in the frontal variant of frontotemporal dementia', *Neuropsychopharmacology*, advanced online publication 7 September 2005, doi:10.1038/sj.npp.1300886

Randall, D. C., Fleck, N. L., Shneerson, J. M. and File, S. E. (2004) 'The cognitive-enhancing properties of modafinil are limited in non-sleep-deprived middle-aged volunteers', *Pharmacology Biochemistry and Behaviour*, vol 77, pp547–555

Rawls, J. A. (1971) *Theory of Justice*, Harvard University Press, Cambridge, MA

Raymond, E. S. (2001) *The Cathedral and the Bazaar*, O'Reilly, Sebastopol, CA

Reardon, D. (2005) 'Unenhanced ethics', *PLoS Medicine*, http://medicine.plosjournals.org/perlserv/?request=read-response&doi=10.1371/journal.pmed.0020121, accessed 24 November 2008

Reginelli, A. D., Wang, Y. Q., Sassoon, D. and Muneoka, K. (1995) 'Digit tip regeneration correlates with regions of Msx1 (Hox 7) expression in fetal and newborn mice', *Development*, vol 121, pp1065–1076

Reing, J. E., Li Zhang, L., Myers-Irvin, J., Cordero, K. E., Freytes, D. O., Heber-Katz, E., Bedelbaeva, K., McIntosh, D., Dewilde, A., Braunhut, S. J. and Badylak, S. F. (in press) 'Degradation products of extracellular matrix affect cell migration and proliferation', *Tissue Engineering*

Repesh, L. A. and Oberpriller, J. C. (1980) 'Ultrastructural studies on migrating epidermal cells during the wound healing stage of regeneration in the adult newt, *Notophthalmus viridescens*', *American Journal of Anatomy*, vol 159, pp187–208

Ressler, K. J., Rothbaum, B. O., Tannenbaum, L., Anderson, P., Graap, K., Zimand, E., Hodges, L. and Davis, M. (2004) 'Cognitive enhancers as adjuncts to psychotherapy: – Use of D-cycloserine in phobic individuals to facilitate extinction of fear', *Archives of General Psychiatry*, vol 61, no 11, pp1136–1144

Rhodes, B. J. and Starner, T. (1996) 'Remembrance agent: A continuously running automated information retrieval system', paper read at the First International Conference on the Practical Application of Intelligent Agents and Multi Agent Technology (PAAM 96), London

Rip, A. (2006) 'A co-evolutionary approach to reflexive governance – and its ironies', in J-P. Voss, D. Bauknecht and R. Kemp (eds) *Reflexive Governance for Sustainable Development*, Edward Elgar, Cheltenham, UK, pp82–100

Rip, A. (2007) 'Die Verzahnung von technologischen und sozialen Determinismen und die Ambivalenzen von Handlungsträgerschaft', in 'Constructive Technology Assessment', in U. Dolata and R. Werle (eds) *Gesellschaft und die Macht der Technik: Sozioökonomischer und Institutioneller Wandel durch Technisierung*, Campus, Frankfurt, Germany, pp83–104

Rip, A. and Groen, A. (2001) 'Many visible hands', in R. Coombs, K. Green, V. Walsh and A. Richards (eds) *Technology and the Market: Demands, Users and Innovation*, Edward Elgar, Cheltenham, UK, pp12–37

Rip, A. and te Kulve, H. (2008) 'Sociotechnical scenarios to support reflexive co-evolution: The approach of Constructive TA', in Center for Nanotechnology in Society (ed) *Yearbook of Nanotechnology in Society*, Arizona State University, Springer, AZ

Rip, A., Misa, T. J. and Schot, J. (eds) (1995) *Managing Technology in Society. The Approach of Constructive Technology Assessment*, Pinter Publishers, London and New York

Robbins, T. W., McAlonan, G., Muir, J. L. and Everitt, B. J. (1997) 'Cognitive enhancers in theory and practice: Studies of the cholinergic hypothesis of cognitive deficits in Alzheimer's disease', *Behavioural Brain Research*, vol 83, pp15–23

Roco, M. and Bainbridge, W. (eds) (2002) *Converging Technologies for Improving Human Performance: Nanotechnology, Biotechnology, Information Technology and Cognitive Science*, US National Science Foundation/Department of Commerce sponsored report, also published by Kluwer Academic Publishers, Dordrecht, The Netherlands

Rogers, R. D., Blackshaw, A. J., Middleton, H. C., Matthews, K., Hawtin, K., Crowley, C., Hopwood, A., Wallace, C., Deakin, J. F., Sahakian, B. J. and Robbins, T. W. (1999) 'Tryptophan depletion impairs stimulus–reward learning while methylphenidate disrupts attentional control in healthy young adults: Implications for the monoaminergic basis of impulsive behaviour', *Psychopharmacology (Berl)*, vol 146, pp482–491

Romanyshyn, R. (1989) *Technology as Symptom and Dream*, Routledge, New York

Ross, J. (2001) 'The organic farmer's story', *The Scotsman*, March 11

Routtenberg, A., Cantallops, I., Zaffuto, S., Serrano, P. and Namgung, U. (2000) 'Enhanced learning after genetic overexpression of a brain growth protein', *Proceedings of the National Academy of Sciences of the USA*, vol 97, no 13, pp7657–7662

Royal Society/Royal Academy of Engineering (2004) 'Nanoscience and nanotechnologies: Opportunities and uncertainties', available at www.nanotec.org.uk/report/contents.pdf, accessed 23 May 2008

Rugino, T. A. and Copley, T. C. (2001) 'Effects of modafinil in children with attention-deficit/hyperactivity disorder: An open-label study', *Journal of the American Academy of Child and Adolescent Psychiatry*, vol 40, pp230–235

Rusted, J. M., Trawley, S., Heath, J., Kettle, G. and Walker, H. (2005) 'Nicotine improves memory for delayed intentions', *Psychopharmacology (Berl)*, vol 182, no 3, pp355–365

Sagarin, E. (ed) (1971) *The Other Minorities: Nonethnic Collectivities Conceptualised as Minority Groups*, Ginn and Co, Toronto, Canada

Salkever, D. S. (1995) 'Updated estimates of earnings benefits from reduced exposure of children to environmental lead', *Environmental Research*, vol 70, no 1, pp1–6

Samulewicz, S. J., Clark, L., Seitz, A. and Heber-Katz, E. (2002) 'Expression of Pref-1, a delta-like protein, in healing mouse ears', *Wound Repair and Regeneration*, vol 10, pp215–221

Sandberg, A. (2001) 'Morphological freedom: Why we not just want it, but *need* it', www.nada.kth.se/~asa/Texts/MorphologicalFreedom.htm, accessed 23 June 2008

Sandberg, A. (2003) 'Morphologic freedom', in *Eudoxa Policy Studies*, available from www.eudoxa.se/content/archives/2003/10/eudoxa_policy_s_3.html, accessed 18 June 2008

Savolainen, P., Zhang, Y. P. et al (2002) 'Genetic evidence for an East Asian origin of domestic dogs', *Science*, vol 298, pp1610–1613

Savulescu, J. (2005) 'New breeds of humans: The moral obligation to enhance', *Ethics, Science and Moral Philosophy of Assisted Human Reproduction*, vol 10, supplement 1, pp36–39

Savulescu, J. (2006) 'Genetic interventions and the ethics of enhancement of human beings', in B. Steinbock (ed) *The Oxford Handbook on Bioethics*, Oxford University Press, Oxford, UK, pp516–535

Savulescu, J. (2007) 'In defence of procreative beneficence: Response to Parker', *Journal of Medical Ethics*, vol 33, pp284–288

Schillerstrom, J. E., Horton, M. S. and Royall, D. R. (2005) 'The impact of medical illness on executive function', *Psychosomatics*, vol 46, no 6, pp508–516

Schneider, J. S., Lee, M. H., Anderson, D. W., Zuck, L. and Lidsky, T. I. (2001) 'Enriched environment during development is protective against lead-induced neurotoxicity', *Brain Research*, vol 896, nos 1–2, pp48–55

Schot, J. and Rip, A. (1997) 'The past and future of constructive technology assessment', *Technological Forecasting and Social Change*, vol 54, pp251–268

Seitz, A., Aglow, E. and Heber-Katz, E. (2002) 'Recovery from spinal cord injury: A new transection model in the C57Bl/6 mouse', *Journal of Neuroscience Research*, vol 67, pp337–345

Sellen, A. J., Louie, G., Harris, J. E. and Wilkins, A. J. (1996) 'What brings intentions to mind? An in situ study of prospective memory', Rank Xerox Research Centre Technical Report EPC-1996-104, Cambridge, UK

Serres, M. with Latour, B. (1998) *Conversations on Science, Culture and Time*, University of Michigan Press, Ann Arbor, MI

Sharpe, B. (2006) 'Applications and impact', in R. G. M. Morris, L. Tarassenko and M. Kenward (eds) *Cognitive Systems: Information Processing Meets Brain Science*, Academic Press, London

Shenoy, K. V., Meeker, D., Cao, S. Y., Kureshi, S. A. , Pesaran, B., Buneo, C. A., Batista, A. R., Mitra, P. P., Burdick, J. W. and Andersen, R. A. (2003) 'Neural prosthetic control signals from plan activity', *Neuroreport*, vol 14, no 4, pp591–596

Sinclair, D. and Guarente, L. (2006) 'Unlocking the secrets of longevity genes', *Scientific American*, March, pp48–57

Singer, P. (2000) *A Darwinian Left: Politics, Evolution, and Cooperation*, Yale University Press, New Haven, CT

Singletary, B. A. and Starner, T. (2000) 'Symbiotic interfaces for wearable face recognition', paper read at HCII2001 Workshop on Wearable Computing, New Orleans, LA

Slattery, D. A., Hudson, A. L. and Nutt, D. J. (2004) 'Invited review: The evolution of antidepressant mechanisms', *Fundamental and Clinical Pharmacology*, vol 18, pp1–21

Smirnoff, N. (2003) 'Vitamin C booster', *Nature Biotechnology*, vol 21, p134–136

Smith, A., Brice, C., Nash, J., Rich, N. and Nutt, D. J. (2003) 'Caffeine and central noradrenaline: Effects on mood, cognitive performance, eye movements and cardiovascular function', *Journal of Psychopharmacology*, vol 17, no 3, pp283–292

Smith, B. D. (2001) 'Documenting plant domestication: The consilience of biological and archaeological approaches', *Proceedings of the National Academy of Sciences of the USA*, vol 98, pp1324–1326

Soetens, E., Casaer, S., Dhooge, R. and Hueting, J. E. (1995) 'Effect of amphetamine on long-term retention of verbal material', *Psychopharmacology*, vol 119, no 2, pp155–162

Soetens, E., Dhooge, R. and Hueting, J. E. (1993) 'Amphetamine enhances human-memory consolidation', *Neuroscience Letters*, vol 161, no 1, pp9–12

Stahl, S. M. (2002a) 'Awakening to the psychopharmacology of sleep and arousal: Novel neurotransmitters and wake-promoting drugs', *Journal of Clinical Psychiatry*, vol 63, pp467–468

Stahl, S. M. (2002b) 'Psychopharmacology of wakefulness: Pathways and neurotransmitters', *Journal of Clinical Psychiatry*, vol 63, pp551–552

Stocum, D. L. and Dearlove, G. E. (1972) 'Epidermal–mesodermal interaction during morphogenesis of the limb regeneration blastema in larval salamanders', *Journal of Experimental Zoology*, vol 181, pp49–62

Stocum, D. L. and Crawford, K. (1987) 'Use of retinoids to analyze the cellular basis of positional memory in regenerating amphibian limbs', *Biochemistry and Cell Biology*, vol 65, pp750–761

Stuss, D. T. and Levine, B. (2002) 'Adult clinical neuropsychology: Lessons from studies of the frontal lobes', *Annual Review of Psychology*, vol 53, pp401–433

Surowiecki, J. (2004) *The Wisdom of Crowds: Why the Many Are Smarter Than the Few and How Collective Wisdom Shapes Business, Economies, Societies and Nations*, Doubleday, New York, NY

Swanson, J. M., Greenhill, L. L., Lopez, F. A., Sedillo, A., Earl, C. Q., Jiang, J. G. and Biederman, J. (2006) 'Modafinil film-coated tablets in children and adolescents with attention-deficit/hyperactivity disorder: Results of a randomized, double-blind, placebo-controlled, fixed-dose study followed by abrupt discontinuation', *Journal of Clinical Psychiatry*, vol 67, pp137–147

Swierstra, T. and Rip, A. (2007) 'Nano-ethics as NEST-ethics: Patterns of moral argumentation about new and emerging science and technology', *NanoEthics*, vol 1, pp3–20

Szeszko, P. R., Bilder, R. M., Dunlop, J. A., Walder, D. J. and Lieberman, J. A. (1999) 'Longitudinal assessment of methylphenidate effects on oral word production and symptoms in first-episode schizophrenia at acute and stabilized phases', *Biological Psychiatry*, vol 45, pp680–686

Tabah, L. (1989) 'From one demographic transition to another', *Population Bulletin of the United Nations*, vol 28, pp1–24

Tannock, R., Schachar, R. J., Carr, R. P., Chajczyk, D. and Logan, G. D. (1989) 'Effects of methylphenidate on inhibitory control in hyperactive children', *Journal of Abnormal Child Psychology*, vol 17, pp473–491

Tassava, R. A., Johnson-Wint, B. and Gross, J. (1986) 'Regenerate epithelium and skin glands of the adult newt react to the same monoclonal antibody', *Journal of Experimental Zoology*, vol 239, pp229–240

Tassava, R. A., Nace, J. D. and Wei, Y. (1996) 'Extracellular matrix protein turnover during salamander limb regeneration', *Wound Repair and Regeneration*, vol 4, no 1, pp75–81

Tatar, M. et al (2003) 'The endocrine regulation of aging by insulin-like signals', *Science*, vol 299, pp1346–1351

Taylor, F. B. and Russo J. (2000) 'Efficacy of modafinil compared to dextroamphetamine for the treatment of attention deficit hyperactivity disorder in adults', *Journal of Child and Adolescent Psychopharmacology*, vol 10, no 4, pp311–320

Taylor, D. A., Atkins, B. Z., Hungspreugs, P., Jones, T. R., Reedy, M. C., Hutcheson, K. A., Glower, D. D. and Kraus, W. E. (1998) 'Regenerating functional myocardium: Improved performance after skeletal myoblast transplantation', *Nature Medicine*, vol 4, pp929–933

Teter, C. J., McCabe, S. E., Cranford, J. A., Boyd, C. J. and Guthrie, S. K. (2005) 'Prevalence and motives for illicit use of prescription stimulants in an undergraduate student sample', *Journal of American College Health*, vol 53, pp253–262

The International Social Survey Program (1998) *Religion II*, German Social Science Infrastructure Services, Cologne, Germany

Thomas, L. (2006) 'Becoming an evil society: The self and strangers', *Political Theory*, vol 24, no 2, pp271–294

Thomson, J. A., Itskovitz-Eldor, J. et al (1998) 'Embryonic stem cell lines derived from human blastocysts', *Science*, vol 282, pp1145–1147

Tieges, Z., Ridderinkhof, R. K., Snel, J. and Kok, A. (2004) 'Caffeine strengthens action monitoring: Evidence from the error-related negativity', *Cognitive Brain Research*, vol 21, no 1, pp87–93

Tomporowski, P. D. (2003) 'Effects of acute bouts of exercise on cognition', *Acta Psychologica*, vol 112, no 3, pp297–324

Turner, D. C. (2005) 'Psychopharmacology of cognitive enhancement', PhD thesis, University of Cambridge, Cambridge, UK

Turner, D. C. (2006) 'A review of the use of modafinil for attention deficit hyperactivity disorder (ADHD)', *Expert Review of Neurotherapeutics*, vol 6, pp455–468

Turner, D. C. and Sahakian, B. J. (2006) 'Analysis of the cognitive enhancing effects of modafinil in schizophrenia', *Progress in Neurotherapeutics and Neuropsychopharmacology*, vol 1, pp133–147

Turner, D. C., Robbins, T. W., Clark, L., Aron, A. R., Dowson, J. and Sahakian, B. J. (2003a) 'Cognitive enhancing effects of modafinil in healthy volunteers', *Psychopharmacology (Berl)*, vol 165, pp260–269

Turner, D. C., Robbins, T. W., Clark, L., Aron, A. R., Dowson, J. and Sahakian, B. J. (2003b) 'Relative lack of cognitive effects of methylphenidate in elderly male volunteers', *Psychopharmacology (Berl)*, vol 168, pp455–464

Turner, D. C., Clark, L., Dowson, J., Robbins, T. W. and Sahakian, B. J. (2004a) 'Modafinil improves cognition and response inhibition in adult attention-deficit/hyperactivity disorder', *Biological Psychiatry*, vol 55, pp1031–1040

Turner, D. C., Clark, L., Pomarol-Clotet, E., McKenna, P., Robbins, T. W. and Sahakian, B. J. (2004b) 'Modafinil improves cognition and attentional set shifting in patients with chronic schizophrenia', *Neuropsychopharmacology*, vol 29, pp1363–1373

Turner, D. C., Blackwell, A. D., Dowson, J. H., McLean, A. and Sahakian, B. J. (2005) 'Neurocognitive effects of methylphenidate in adult attention-deficit/hyperactivity disorder', *Psychopharmacology (Berl)*, vol 178, pp286–295

United Nations (1948) 'UN Declaration on Human Rights', www.un.org/Overview/rights.html, accessed 18 June, 2008

United Nations (1995) 'Final resolution', UN World Summit on Social Development, Copenhagen, United Nations, New York

United Nations (1999) *Long-Range World Population Projections: Based on the 1998 Revision*, ESA/P/WP.153, United Nations, New York

United Nations (2003) *World Population Prospects: The 2002 Revision*, United Nations, New York

United Nations General Assembly (1998) 'Universal Declaration on the Human Genome and Human Rights', UN General Assembly, United Nations, New York

Vaillant, G. E. (2002) *Aging Well: Surprising Guideposts to a Happier Life from the Landmark Harvard Study of Adult Development*, Little, Brown and Company, London

van Beek, T. A. (2002) 'Chemical analysis of Ginkgo biloba leaves and extracts', *Journal of Chromatography A*, vol 967, no 1, pp21–55

van Laere, A. S., Nguyen, M. et al (2003) 'A regulatory mutation in IGF2 causes a major QTL effect on muscle growth in the pig', *Nature*, vol 425, pp832–836

Vaynman, S. and Gomez-Pinilla, F. (2005) 'License to run: Exercise impacts functional plasticity in the intact and injured central nervous system by using neurotrophins', *Neurorehabilitation and Neural Repair*, vol 19, no 4, pp283–295

Vergara, M. et al (2004) 'Hormone-treated Snell dwarf mice regain fertility but remain long-lived and disease resistant', *Journals of Gerontology Series A – Biological and Medical Sciences*, vol 59, pp1244–1250

von Wild, K., Rabischong, P., Brunelli, G., Benichou, M. and Krishnan K. (2002) 'Computer added locomotion by implanted electrical stimulation in paraplegic patients (SUAW)', *Acta Neurochirurgica Supplement*, vol 79, pp99–104

Walsh, R. N., Budtz-Olsen, O. E., Penny, J. E. and Cummins, R. A. (1969) 'The effects of environmental complexity on the histology of the rat hippocampus', *The Journal of Comparative Neurology*, vol 137, no 3, pp361–365

Wang, H. B., Ferguson, G. D., Pineda, V. V., Cundiff, P. E. and Storm, D. R. (2004) 'Overexpression of type-1 adenylyl cyclase in mouse forebrain enhances recognition memory and LTP', *Nature Neuroscience*, vol 7, no 6, pp635–642

Wang, L., Zhang, X. and Coady, D. (2002) 'Health inequality and its causes: An empirical analysis based on the first China children health survey', in Y. Yang (ed) *Equity and Social Equality in Transitional China*, Renmin University Press, Beijing, pp229–253

Warburton, D. M. (1992) 'Nicotine as a cognitive enhancer', *Progress in Neuropsychopharmacology and Biological Psychiatry*, vol 16, no 2, pp181–191

Warner, H. (2005) *Twenty Years of Progress in Biogerontology*, National Institute on Aging, Bethesda, MD

Warner, H., Anderson, J., Austad, S., Bergamini, E., Bredesen, D., Butler, R., Carnes, B. A., Clark, B. F., Cristofalo, V., Faulkner, J., Guarente, L., Harrison, D. E., Kirkwood, T., Lithgow, G., Martin, G., Masoro, E., Melov, S., Miller, R. A., Olshansky, S. J., Partridge, L., Pereira-Smith, O., Perls, T., Richardson, A., Smith, J., von Zglinicki, T., Wang, E., Wei, J. Y. and Williams, T. F. (2005) 'Science fact and the SENS agenda: What can we reasonably expect from ageing research?', *EMBO Reports*, vol 6, no 11, pp1006–1008

Warwick, K. (2002) *I, Cyborg*, Century, Post Falls, ID

Warwick, K., Gasson, M., Hutt, B., Goodhew, I., Kyberd, P., Andrews, B., Teddy, P. and Shad, A. (2003) 'The application of implant technology for cybernetic systems', *Archives of Neurology*, vol 60, no 10, pp1369–1373

Warwick, K., Gasson, M., Hutt, B., Goodhew, I., Kyberd, P., Schulzrinne, H. and Wu, X. (2004) 'Thought communication and control: A first step using radiotelegraphy', *IEE Proceedings on Communications*, vol 151, no 3, pp185–189

Wei, F., Wang, G. D., Kerchner, G. A., Kim, S. J., Xu, H. M., Chen, Z. F. and Zhuo, M. (2001) 'Genetic enhancement of inflammatory pain by forebrain NR2B overexpression', *Nature Neuroscience*, vol 4, no 2, pp164–169

Weindruch, R. and Sohal, R. S. (1997) 'Seminars in medicine of the Beth Israel Deaconess Medical Center: Caloric intake and aging', *New England Journal of Medicine*, vol 337, pp986–994

Weindruch, R. and Walford, R. L. (1988) *The Retardation of Aging and Disease by Dietary Restriction*, Charles C. Thomas, Springfield, IL

Weiser, M. (1991) 'The computer for the twenty-first century', *Scientific American*, vol 265, no 3, pp94–110

Wenk, G. L. (1989) 'An hypothesis on the role of glucose in the mechanism of action of cognitive enhancers', *Psychopharmacology*, vol 99, pp431–438

Whalley, L. J. and Deary, I. J. (2001) 'Longitudinal cohort study of childhood IQ and survival up to age 76', *British Medical Journal*, vol 322, no 7290, pp819–822

Whitesides, G. M. (2004) 'Assumptions: Taking chemistry in new directions', *Angewandte Chemie International Edition*, vol 43, pp3632–3641

WHO (2005) 'The Bangkok Charter for Health Promotion in a Globalized World', 6th Global Health Promotion Conference, Bangkok, www.who.int/healthpromotion/conferences/6gchp/bangkok_charter/en/print.html and www.who.int/healthpromotion/conferences/6gchp/hpr_050829_%20BCHP.pdf, accessed 18 June 2008

Wiggins, J. S. (ed) (1996) *The Five-Factor Model of Personality*, Guilford, New York

Wilkinson, L., Scholey, A. and Wesnes, K. (2002) 'Chewing gum selectively improves aspects of memory in healthy volunteers', *Appetite*, vol 38, no 3, pp235–236

Wilmut, I., Schnieke, A. E. et al (1997) 'Viable offspring derived from fetal and adult mammalian cells', *Nature*, vol 385, pp810–813

Winder, R. and Borrill, J. (1998) 'Fuels for memory: The role of oxygen and glucose in memory enhancement', *Psychopharmacology*, vol 136, no 4, pp349–356

Winship, C. and Korenman, S. (1997) 'Does staying in school make you smarter? The effect of education on IQ in *The Bell Curve*', in B. Devlin, S. E. Fienberg and K. Roeder (eds) *Intelligence, Genes, and Success: Scientists Respond to* The Bell Curve, Springer, New York, pp215–234

Wolbring, G. (2004) 'Solutions follow perception: Nano-Bio-Info-Cogno-technology (NBIC) and the concept of health, medicine, disability and disease', *Alberta Health Law Review*, vol 12, no 3, pp41–47, www.law.ualberta.ca/centres/hli/pdfs/hlr/v12_3/12-3-10%20Wolbring.pdf, accessed 18 June 2008

Wolbring, G. (2005) *The Triangle of Enhancement Medicine, Disabled People, and the Concept of Health: A New Challenge for HTA, Health Research, and Health Policy*, HTA Initiative No 23, ISBN 978-1-894927-36-9 (print), ISBN 978-1-894927-37-6 (online), ISSN 1706-7855, www.ihe.ca/documents/hta/HTA-FR23.pdf, accessed 18 June 2008

Wolbring, G. (2006a) 'Synthetic biology 2.0', *The Choice is Yours*cColumn, www.innovationwatch.com/choiceisyours/choiceisyours.2006.05.30.htm, accessed 18 June 2008

Wolbring, G. (2006b) 'Human security and NBICS', www.innovationwatch.com/choiceisyours/choiceisyours.2006.12.30.htm, accessed 18 June 2008

Wolbring, G. (2006c) 'Ableism and NBICS', www.innovationwatch.com/choiceisyours/choiceisyours.2006.08.15.htm, accessed 18 June 2008

Wolbring, G. (2006d) 'Three challenges to the Ottawa spirit of health promotion, trends in global health, and disabled people', *Canadian Journal of Public Health*, vol 97, no 5, pp405–408

Wolbring, G. (2006e) 'The unenhanced underclass', in P. Miller and J. Wilsdon (eds) *Better Humans? The Politics of Human Enhancement*, Demos, London, pp122–128

Wolbring, G. (2007a) 'Synthetic biology 3.0', www.innovationwatch.com/choiceisyours/choiceisyours-2007-07-15.htm, accessed 18 June 2008

Wolbring, G. (2007b) 'Enhancement of animals', www.innovationwatch.com/choiceisyours/choiceisyours-2007-03-15.htm, accessed 18 June 2008

Wolbring, G. (2007c) 'New and emerging sciences and technologies, ableism, transhumanism and religion, faith, theology and churches', *Madang: International Journal of Contextual Theology in East Asia*, vol 7, pp79–112

Wolbring, G. (2007d) 'NBICS, other convergences, ableism and the culture of peace', Innovationwatch.com, www.innovationwatch.com/choiceisyours/choiceisyours-2007-04-15.htm, accessed 18 June 2008

Wolbring, G. (2007e) 'What convergence is in the cards for future scientists?', conference presentation, Vienna, May, hosted on International Center for Bioethics Culture and Disability webpage, http://bioethicsanddisability.org/convergence.htm, accessed 18 June 2008

Wolbring, G. (2007f) 'NBICS and social cohesion', www.innovationwatch.com/choiceisyours/choiceisyours.2007.01.15.htm, accessed 18 June 2008

Wolbring, G. (2007g) 'From NBICS to ABECS: From S&T-convergence to skill, issue and stakeholder convergence to ism convergence innovation', *The European Journal of Social Science Research* (submitted)

Wolbring, G. (2007h) 'Glossary for the 21st century', International Center for Bioethics, Culture and Disability webpage, www.bioethicsanddisability.org/glossary.htm, accessed 18 June 2008

Wolbring, G. (2007i) 'Robotics, artificial intelligence, sentient rights, speciesism, and uploading the mind', www.innovationwatch.com/choiceisyours/choiceisyours-2007-02-15.htm, accessed 18 June, 2008

Wolfensohn, J. D. (2002) 'Poor, disabled and shut out', *Washington Post*, 3 December, pA25

Wolpaw, J., McFarland, D., Neat, G. and Forneris, C. (1991) 'An EEG-based brain–computer interface for cursor control', *Electroencephalography and Clinical Neurophysiology*, vol 78, no 3, pp252–259

Wolpaw, J. R., Birbaumer, N., Heetderks, W. J., McFarland, D. J., Peckham, P. H., Schalk, G., Donchin, E., Quatrano, L. A., Robinson, C. J. and Vaughan, T. M. (2000) 'Brain–computer interface technology: A review of the first international meeting', *IEEE Transactions on Rehabilitation Engineering*, vol 8, no 2, pp164–173

World Transhumanist Association (2002) 'The Transhumanist Declaration', www.transhumanism.org/index.php/WTA/declaration, accessed 2 June 2008

World Transhumanist Association (2003) 'Transhumanist FAQ – A general introduction – Version 2.1', www.transhumanism.org/resources/faq, accessed 23 June, 2008

Yang, E. V. and Bryant, S. V. (1994) 'Developmental regulation of a matrix metalloproteinase during regeneration of axolotl appendages', *Developmental Biology*, vol 166, pp696–703

Yang, K., Marsh, T., Mun, M. and Shahabi, C. (2005) 'Continuous archival and analysis of user data in virtual and immersive game environments', *Proceedings of the ACM Multimedia Conference*, Association for Computing Machinery, New York

Yates, F. (1966) *The Art of Memory*, University of Chicago Press, Chicago, IL

Yerkes, R. M. and Dodson, J. D. (1908) 'The relation of strength of stimulus to rapidity of habit-formation', *Journal of Comparative Neurology and Psychology*, vol 18, pp459–482

Yu, B. P. et al (1985) 'Nutritional influences on aging of Fischer 344 rats: I. Physical, metabolic, and longevity characteristics', *Journals of Gerontology*, vol 40, pp657–670

Yu, H., Mohan, S., Masinde, G. L. and Baylink, D. J. (2005) 'Mapping the dominant wound healing and soft tissue regeneration QTL in MRL x CAST', *Mammalian Genome*, vol 16, pp918–924

Yu, N., Chen, J. and Ju, M. (2001) 'Closed-loop control of quadriceps/hamstring activation for FES-induced standing-up movement of paraplegics', *Journal of Musculoskeletal Research*, vol 5, no 3, pp173–184

Zohlnhöfer, D., Ott, I., Mehilli, J., Schömig, K., Michalk, F., Ibrahim, T., Meisetschläger, G., von Wedel, J., Bollwein, H., Seyfarth, M., Dirschinger, J., Schmitt, C., Schwaiger, M., Kastrati, A. and Schömig, A. (2006) 'Stem cell mobilization by granulocyte colony-stimulating factor in patients with acute myocardial infarction, a randomized controlled trial', *Journal of the American Medical Association*, vol 295, pp1003–1010

Zheng, J. et al (2005) 'Hepatitis B vaccine coverage rate and influential factors among children in Hubei Province', *Journal of Public Health and Preventive Medicine*, vol 16, no 1

Zhu, X. et al (1998) 'The hepatitis B (HB) vaccine coverage rate in 10 Chinese provinces and its influence factors', *Immunization in China*, vol 4, no 4

Zlatos, B. (2004) 'Pitt home to promising study on the brain-dead', *Pittsburgh Tribune-Review*, 27 June, www.pittsburghlive.com/x/pittsburghtrib/s_200768.html, accessed 21 May 2008

Index